John Jennings Moorman

Mineral Springs of North America

How to Reach and How to Use Them

John Jennings Moorman

Mineral Springs of North America
How to Reach and How to Use Them

ISBN/EAN: 9783337371852

Printed in Europe, USA, Canada, Australia, Japan

Cover: Foto ©berggeist007 / pixelio.de

More available books at **www.hansebooks.com**

MINERAL SPRINGS

OF

NORTH AMERICA;

HOW TO REACH, AND HOW TO USE THEM.

BY

J. J. MOORMAN, M.D.,

Physician to the White Sulphur Springs; Professor of Medical Jurisprudence and Hygiene in Washington University, Baltimore; Member of the Medico-Chirurgical Society of Maryland; of the Baltimore Medical Association, etc.

PHILADELPHIA:
J. B. LIPPINCOTT & CO.
1873.

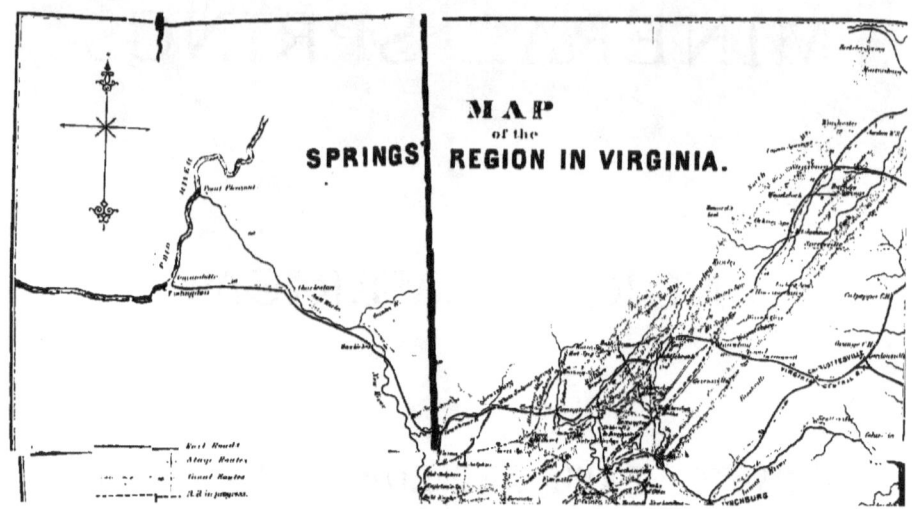

TO

THOSE FOR WHOM I HAVE PRESCRIBED MINERAL WATERS FOR THE LAST THIRTY-FIVE YEARS,

THIS VOLUME,

INTENDED TO DEFINE THE CHARACTER AND APPLICABILITIES OF THE MINERAL WATERS OF AMERICA, IS RESPECTFULLY INSCRIBED BY

THE AUTHOR.

TO THE PUBLIC.

For more than thirty-five years I have given special attention to the investigation of the nature and medicinal applicability of mineral waters. During this time I have resided, throughout the watering season, at the White Sulphur Springs, where, in the character of physician to the springs, I have enjoyed ample opportunities of witnessing the various and modified effects of the water in almost every variety of disease and state of the system.

Although my attention, during this time, has been particularly directed to the investigation of the character of the water of that spring, I have not neglected the other valuable waters of the country, nor failed personally to observe and appreciate their various peculiarities, and their relative and positive merits.

While my position has enabled me to witness the virtues of mineral waters in diseases, it has, at the same time, fully satisfied me not only that their good effects are often lost, but that consequences highly injurious frequently result from their injudicious use.

Impressed with the importance of arresting the abuse of the White Sulphur waters, and of leading to a more correct administration of them, I published, in 1839, a pamphlet designed as a "Directory" for the use of these waters. It was with diffidence I undertook this pioneer effort in a field so entirely unexplored; for, although thousands of invalids had, for more than half a century, annually resorted to these waters, up to the period of issuing the "Directory" not a line had ever been published relative to their medicinal applicability, or the proper methods of prescribing them.

Satisfied from experience that the little *effort* alluded to was not without beneficial effects in guiding to a more prudent use of the waters, I published, in 1846, a small volume entitled "Virginia Springs," designed to embrace what was then known of the various mineral springs in Virginia.

In 1855, and again in 1857, new and enlarged editions of the work were issued. In 1859, the previous editions having been exhausted, a new one, much enlarged, and embracing not only the Virginia Springs, but also the springs of the Southern and Western States, was issued, under the title of the "*Virginia Springs and Springs of the South and West.*" This was followed, in 1867, by a larger and more comprehensive volume, entitled "*Mineral Springs of the United States and Canada.*" Since the publication of that work, important mineral waters in various parts of the United States have come into practical use; some possessing medicinal applicabilities of admitted value, and many claiming valuable therapeutic powers that make them worthy of general notice. These facts, in connection with the obviously growing importance in the public mind of mineral waters generally as remedial agents, and the suggestion of many kind friends, induce me to bring out the present volume, under the title of "*Mineral Springs of North America.*"

A gratifying public appreciation and generous demand for my previous volumes, encourage me to hope that the present one will be an acceptable addition to our very limited spring literature.

In a notice so extensive of mineral fountains, with the exception of those of which I have a personal knowledge, I have necessarily had to depend largely upon the observations and writings of others; and, in this connection, I desire to express my obligations especially to the labors of my esteemed friend, Dr. Bell, of Philadelphia, from whose works and correspondence I have derived important facilities.

In treating of springs as medicinal agents (and it is in that point of view only that I have proposed to treat of them), it has been my earnest effort to present them before the public in an aspect as full and impartial as was possible. So far as the author's personal knowledge and experience, or reliable information obtained from other sources, have enabled him to do so, he has discharged the task with fidelity.

In some instances reliable analyses have not been made of some mineral fountains whose rising importance deserves such chemical test. Nor have these fountains, as yet, furnished, from observation, such record of their adaptations as is desirable in forming a proper appreciation of their merits; hence, in reference to the precise quality and adaptations of such springs, we are necessarily left to inferences based upon analogies and somewhat uncertain comparisons.

The absence of an analysis of a mineral water is less to be regretted if a fair and reliable record of its virtues and appropriate medical uses be obtained; for it is only by multiplied facts, that is, by *experience of its use*, that we can speak positively of its effects. This being so, it is of especial importance that there should be an intelligent resident physician at each fountain, who would make it his duty carefully to note the character of the various diseases submitted to its use, and the effects of the water upon each case. Under such a system, each fountain would soon establish a reliable *record* for itself; the invalid would be greatly assisted in his selection of the proper agent to which he should resort, and the just character of each water be properly understood, and placed upon a firm and stable foundation. This field of observation offers large and exciting motives to a proper medical ambition; for such, as a general thing, has hitherto been the wild and haphazard empiricism in the use of mineral waters in America, and such is the importance of so classifying and systemizing their uses that they may be prescribed understandingly and safely, that he who may contribute to this end, and thus render them the safe, certain, and effective remedies they were designed to be by a beneficent Providence, may well feel that he has neither lived nor labored in vain in his generation.

I will only add that I have endeavored, in getting up this work, to adhere to the plain, unassuming, practical method, which was, I think, a characteristic distinction of my previous volumes, and perhaps their chief merit.

It has been my earnest desire to place in the hands of the public, and especially of invalids, a short and easy, but a condensed and comprehensive, account of the mineral springs of the American continent, and to indicate with candor, and with as much plainness as possible, their nature and medicinal applicability.

Wherever I could, with advantage to the public, I have availed myself of the observations of others, and I claim at the hands of my readers this award of merit at least: *of having honestly endeavored to make my humble labors convenient and practically valuable to them;* not by dazzling but uncertain theories, nor by creating hopes that might end in sad disappointment, but by plain, practical facts in relation to the nature and proper uses of our various mineral waters.

In arranging the matter for the volume, I shall treat of the waters under the heads of the States in which they are respectively found; and have preferred to introduce the States rather in the order of their *mineral water* similitudes than in the usual geographical or political order in which they are generally made to stand. Hence I shall first treat of the waters of Virginia and West Virginia, and of the Western and Southern States; and then of those of the North and East, commencing with the great mineral water State of New York.

I have intentionally avoided in this, as in my previous volumes, all criticisms upon the improvements of spring property, or of the character of the accommodations at the several springs. Such criticism, in a printed volume intended for reference long after its issue from the press, would be likely to mislead, and probably do great injustice; inasmuch as improvements, now faulty, may, before the next season, be rendered very comfortable, and bad hotel accommodations are often amended in a day by a change of landlord or manager. It is of the *nature and medicinal applicability* of mineral waters that I have felt called upon to write; and this I have done without prejudice, fear, or favor; having no interest, directly or indirectly, in any of the springs, and influenced alone in my estimation of them by personal observation, or, when this has been wanting, from the most reliable information I could obtain.

I am not vain enough to suppose that none of my opinions are erroneous: to err is both human and common; but upon the honest integrity with which they have been formed, the invalid, the profession, and the general public may rely.

J. J. MOORMAN.

WHITE SULPHUR SPRINGS, W. VA.,
    March, 1873.

CONTENTS.

CHAPTER I.
MINERAL WATERS IN GENERAL.
PAGE
Early Use of, etc.—Experience the only Guide in the Administration—Medical Efficacy—Modus Operandi, etc.—Length of Time to be Used—General Remarks on Administration............ 21

CHAPTER II.
MINERAL WATERS IN GENERAL (CONTINUED).
Resemblance of some Mineral Waters to Mercury—Errors and Abuse of Mineral Waters, etc.—Changing from Spring to Spring—Dress—Diet, Exercise—Best Time for Using—Length of Time to be Used, etc.. 35

CHAPTER III.
USE OF MEDICINES AND DIFFERENT MINERAL WATERS.
Prescribing Mineral Water—The Best Period of the Year for Invalids to Visit the Springs.. 50

CHAPTER IV.
WEST VIRGINIA AND VIRGINIA SPRINGS.
Routes to the West Virginia and Virginia Springs.................... 59

CHAPTER V.
WHITE SULPHUR SPRINGS.
Location and General Physical Characteristics—Its Strength uniformly the same—Does not lose its Strength by parting with its Gas—Does not deposit its Salts when Quiescent—Its Gas fatal to Fish—Its Early History—Known to the Indians as a "Medi-

cine Water"—First used by the Whites in 1778—Progress of Improvements, and present Condition—Analyses of Mr. Hayes and Professor Rogers.. 62

CHAPTER VI.

THE RELATIVE VIRTUES OF THE SALINE AND GASEOUS CONTENTS OF THE WHITE SULPHUR WATER................. 71

CHAPTER VII.

GENERAL DIRECTIONS FOR THE USE OF THE WHITE SULPHUR WATER.

Directions meant to be General, not Specific—Must not look to the Sensible Operations of the Water for its Best Effects—Moderate or Small Quantities Generally Preferable—Necessary Preparations of the System for the Use of the Water—Sensible Medicinal Effects of the Water—Effects on the Pulse—Synopsis of Rules to be Observed—Use of Baths.............................. 81

CHAPTER VIII.

DISEASES IN WHICH THE WHITE SULPHUR WATER MAY, OR MAY NOT, BE USEFULLY PRESCRIBED.

Dyspepsia—Gastralgia—Water-Brash—Chronic Gastro-Enteritis—Diseases of the Liver—Jaundice—Enlargement of the Spleen—Chronic Irritation of the Bowels—Costiveness—Piles—Diseases of the Urinary Organs—Chronic Inflammation of the Kidneys—Diabetes—Female Diseases: Amenorrhœa, Dysmenorrhœa, Chlorosis, Leucorrhœa—Chronic Affections of the Brain—Nervous Diseases—Paralysis—Some Forms of Chronic Diseases of the Chest, or Breast Complaints (to be avoided in Pulmonary Consumption)—Bronchitis—Chronic Diseases of the Skin, Psoriasis, Lepra, Ill-conditioned Ulcers—Rheumatism and Gout—Dropsies—Scrofula—Mercurial Diseases—Erysipelas—Effects in Inebriates—Effects upon the Opium-eaters—Not to be Used in Diseases of the Heart, or in Scirrhus and Cancer—Chalybeate Spring at the White Sulphur.............................. 91

CHAPTER IX.

SALT SULPHUR SPRINGS.

Location, etc.—Analysis by Professor Rogers—Medical Applicability of the Waters.. 109

CONTENTS.

CHAPTER X.
RED SULPHUR SPRINGS.

Location—Analysis—Adaptation to Diseases, etc.—New River White Sulphur Springs.. 111

CHAPTER XI.
SWEET SPRINGS.

Situation and Early History—Improvements—Analysis—Effects of the Waters—Adaptation of the Waters as a Beverage and as a Bath, etc.. 115

CHAPTER XII.
SWEET CHALYBEATE, OR RED SWEET SPRINGS.

Their Analysis—Nature and Medicinal Adaptations of the Waters as a Beverage and a Bath—Artificial Warm Baths, etc.......... 121

CHAPTER XIII.
HOT SPRINGS.

Effects of the Waters Internally and Externally used—Analysis—Diseases to which they are applicable—Speculations on Thermalization, etc.. 128

CHAPTER XIV.
WARM SPRINGS.

Analysis—Time and Manner of Using—Diseases for which Employed, etc.. 134

CHAPTER XV.
HEALING SPRINGS.

Location—Analyses—Therapeutic Action—Diseases for which they may be Prescribed, etc.. 137

CHAPTER XVI.
ROCKBRIDGE ALUM SPRINGS.

Location—Analysis—Remarks on Analysis—The Name Alum a Misnomer, etc.—Therapeutic Effects of the Waters—Diseases in which they are employed—Their Excellent Effects in Scrofula 141
Jordon Rockbridge Alum Springs.. 146

CHAPTER XVII.
BATH ALUM SPRINGS.

	PAGE
Analysis—Diseases and States of the System in which they may be Prescribed, etc.	147

CHAPTER XVIII.

Rockbridge Baths	150
Cold Sulphur Spring	150
Variety Springs	151
Stribling's Springs	151

CHAPTER XIX.

Rawley's Spring	154
Massanetta Springs	156
Jordan's White Sulphur Springs	157

CHAPTER XX.
BATH OR BERKELEY SPRINGS.

Early History—Baths and Bathing-Houses—Medical Properties of the Waters—Diseases for which used, etc.	159
Capon Springs	161

CHAPTER XXI.

Coiner's Black and White Sulphur Springs	163
Roanoke Red Sulphur Spring	164
Johnson's Springs	164
The Blue Ridge Spring	164
Alleghany Springs	165
Montgomery White Sulphur	170

CHAPTER XXII.

Yellow Sulphur Springs	171
Pulaski Alum Spring	174
Grayson Sulphur Springs	174
Holston Springs	175
Kimberling Springs	176

CONTENTS.

CHAPTER XXIII.

	PAGE
Fauquier White Sulphur Springs	178
Buffalo Springs	178
Huguenot Springs	180
New London Alum Spring	181

CHAPTER XXIV.
SPRINGS OF KENTUCKY.

Harrodsburg—Rochester—Olympian—Blue Lick—Estill	183

CHAPTER XXV.
MINERAL SPRINGS OF OHIO AND INDIANA.

Ohio White Sulphur	188
Yellow Springs	189
French Lick Springs, Indiana	190

CHAPTER XXVI.
SPRINGS OF MICHIGAN AND WISCONSIN.

St. Louis Springs, Michigan	192
Bethesda Springs, Wisconsin	194

CHAPTER XXVII.
SPRINGS OF TENNESSEE.

White's Creek Spring—Robertson's—Winchester—Beersheba—Montvale—Tate's—Lee's—Sulphur and Chalybeate—Alum Springs—Warm Springs on the French Broad	195

CHAPTER XXVIII.
SPRINGS OF NORTH CAROLINA.

Warm and Hot Springs of Buncombe—Shocco Springs—Jones' White Sulphur and Chalybeate—Kittrell's Springs	200
Sulphur Springs, Catawba County	203

CHAPTER XXIX.
SPRINGS OF SOUTH CAROLINA.

Glenn's—West's—Springs in Abbeville and Laurens Districts, etc.—Chick's—Williamstown Springs—Artesian Well in Charleston	204

CHAPTER XXX.
SPRINGS OF GEORGIA.
Indian—Madison—Warm Springs—Gordon's—Catoosa Springs   PAGE 206

CHAPTER XXXI.
SPRINGS OF ALABAMA.
Bladen Springs—Bailey's Spring—Tallahatta Springs.............. 208

CHAPTER XXXII.
SPRINGS OF MISSISSIPPI.
Cooper's Well—Ocean Springs.. 216

CHAPTER XXXIII.
SPRINGS OF ARKANSAS.
Washita Hot Springs... 213
Springs of Florida... 217

CHAPTER XXXIV.
MINERAL SPRINGS OF NEW YORK.
Saratoga and Ballston Group—Classification of Waters—Geological Position—Thermalization of Waters—Analysis of various Springs, etc.. 218

CHAPTER XXXV.
NEW YORK MINERAL WATERS (CONTINUED).
Improper Use of the Saratoga Waters, and its Evils—Injurious Advice and Errors of Opinion as to the Nature and Use of Mineral Waters... 229
Diseases for which the Saratoga Waters may be Prescribed—Albany Artesian Well—Reed's Mineral Spring—Halleck's Spring, etc... 231

CHAPTER XXXVI.
NEW YORK SULPHUR SPRINGS.
Sharon Springs—Avon Springs—Richfield Springs................. 238

CONTENTS.

CHAPTER XXXVII.
NEW YORK SULPHUR AND ACIDULOUS SPRINGS.

PAGE

Clifton Springs—Chittenango Springs—Messina Sulphur Springs—Manlius Springs—Auburn Springs—Chappaqua Springs—Harrowgate Spring—Spring at Troy—Newburg Spring—Springs in Dutchess and Columbia Counties—Catskill Spring—Nanticoke Spring—Dryden Spring—Rochester Spring—Springs in Monroe County: Gates, Mendon, and Ogden—Verona Spring—Saquoit Springs—Springs in Niagara County—Seneca or Deer Lick Springs—Oak Orchard Acid Springs—Acid Spring at Clifton—Adirondack Spring...................... 245

CHAPTER XXXVIII.
SPRINGS OF PENNSYLVANIA.

Bedford Springs—Gettysburg Spring—Frankfort Mineral Springs Chalybeate Spring near Pittsburg—York Springs—Carlisle Springs—Perry County Springs—Doubling Gap and Chalybeate Springs—Fayette Spring—Bath Chalybeate Spring—Blossburg Spring—Ephrata Springs—Yellow Springs—Caledonia Springs... 255

CHAPTER XXXIX.
MINERAL SPRINGS OF VERMONT.

Clarendon Gaseous Springs—Newburg Sulphur Springs—Highgate Springs—Abburgh Spring—Missisquoi Springs—Vermont Springs—Alburgh Springs....................................... 268

CHAPTER XL.
SPRINGS OF MASSACHUSETTS.

Hopkinton Springs—Berkshire Soda Spring........................... 272

CHAPTER XLI.
SPRINGS OF NEW JERSEY AND MAINE.

Schooley's Mountain Spring.. 274
Saline Lubec Spring in Maine—Dexter Chalybeate Spring.......... 275

CHAPTER XLII.

MINERAL AND THERMAL WATERS BETWEEN THE MISSISSIPPI AND THE PACIFIC OCEAN.

PAGE

In California — Oregon — Kansas — New Mexico — Wyoming — Utah, etc.. 277

Table exhibiting the Thermalization of the Various Warm and Hot Springs of the United States and its Territories.............. 284

CHAPTER XLIII.

MINERAL SPRINGS OF CANADA.

Caledonia Springs—Charlottesville Spring—St. Catharine's Artesian Wells—Varennes Springs—St. Leon Spring—Plantagenet Spring—Caxton Spring... 286

MINERAL SPRINGS

OF

NORTH AMERICA.

CHAPTER I.

MINERAL WATERS IN GENERAL.

Early Use of, etc.—Experience the only Guide in the Administration—Medical Efficacy—Modus Operandi, etc.—Length of Time to be Used—General Remarks on Administration.

MINERAL waters rank among the ancient remedies used for the cure of disease. The Greeks, who in knowledge of medicine were superior to the nations who had preceded them, regarded natural medicated waters as a special boon of the Deity, and piously dedicated them to Hercules, the god of strength. They used them for drinking, and for general and topical bathing. Hippocrates was acquainted with the value and uses of various mineral waters, and many other Greek physicians, we are told, employed them for numerous diseases for which they are used at this day.

With the Romans, mineral waters were a familiar remedy, not only in Italy, but in all the countries over which that nation obtained dominion. Mineral springs were eagerly sought out in the countries over which their conquests from time to time extended, and prompted by "gratitude for the benefit which they experienced from their use, they decorated them with edifices, and each fount was placed under the protection of a tutelary deity." (*Bell.*) Pliny, in his Natural History, treats of various mineral waters and their uses; and it is a fact worthy of remark, that they were highly recommended by various Roman physicians, in the fifth century, in the same diseases for which they are at this day so much employed,—particularly for nervous and rheumatic diseases, and for derangements of the liver, stomach, and skin.

With the modern nations of civilized Europe, mineral waters, both as internal and external remedies, have always been held in high estimation. The national regulations that have from time to time been adopted to investigate their virtues and their appropriate applicability, and to guard against their improper use, sufficiently manifest the importance that has been attached to them as remedial agents. Henri IV., we are told, "during his youth had frequented the springs of the Pyrenees, and witnessing the abuses in the employment of so useful a remedy, sought to correct them after his accession to the throne of France. He nominated, by edicts and letters-patent, in 1603, superintendents and superintendents-general, who were charged with the entire control over the use of mineral waters, baths, and fountains of the kingdom. Most of the mineral springs and bathing establishments on the continent of Europe are placed under a somewhat similar superintendence, and a resident physician is also appointed by the government." (*Bell.*)

Although mineral waters had been favorite remedial agents with the enlightened nations of the earth for many centuries, it was comparatively but recently that chemistry, by minute analysis, was able to determine with precision their constituent parts.

In 1670, the mineral waters of France were first fully analyzed by a commission appointed by the Academy of Sciences at Paris; but it was not until 1766, nearly a hundred years afterwards, that Bayen discovered the means of separating sulphur from sulphurous waters,— nor until 1774 that the celebrated Bergmann demonstrated the existence of sulphuretted hydrogen gas. Meanwhile, physicians stationed at the several watering-places were active in observing and noting the various operations of the different waters on the human system, and in determining, from experience, the various cases in which they were beneficial or injurious.

Experience the only sure Guide in the Administration, etc.—After all that science can effect in determining the component parts of mineral waters, it is *experience* alone in their use that can be fully relied upon as to their specific effects, or applicability to particular diseases. Chemical analysis is important mainly as a matter of general scientific knowledge, and may be so far practically useful to the physician as to enable him to form correct *general views* as relates to the nature and powers of the remedy; but it is fallacious to suppose that an analysis, however perfect, can ever enable the physician, in the present state of our knowledge, and in the *absence of practical observation*, to prescribe a mineral water with confidence or safety. An accurate knowledge of the component parts of mineral waters might do much, I admit, to prevent the incessant mistakes and mischief which medical men commit in sending their patients, *hap-hazard*, to drink mineral waters which are often unadapted to their cases; but it never can, in the absence of experimental knowledge, qualify them for giving specific and detailed directions for their use. Dr. John Bell, in his valuable work on "Baths and Mineral Waters," has the following sensible and judicious passage upon this subject: "I wish not," he says, "to be ranked among the chemical physicians, who, having discovered the proportion of each foreign ingredient in the mineral spring, and studied its operation on the economy, pretend to determine the general effect of the compound. We may, indeed, by a knowledge of the constituent parts, predict to a certain extent the medicinal power of the compound; but it is only by multiplied facts, that is, *experience of its use*, that we can speak positively of its virtues."

In no other country, perhaps, do mineral waters abound in greater variety than in the United States; and it is a subject of sincere regret, that their nature, applicability, and proper method of administration should have been so little studied, both by physicians and the public at large. It is true that certain opinions generally prevail in enlightened circles as regards the curative powers of some of our more celebrated fountains; and these opinions, so far as they go, being generally founded on experience, may, in the main, be tolerably correct. Nevertheless, there is a lamentable want of information generally, and even among our more enlightened physicians, as to the *specific nature and adaptation of mineral waters to particular diseases*—information the want of which must always disqualify for the safe and confident recommendation of these valuable agents.

A perfect knowledge of the various influences and of the peculiar minute circumstances that control the use of mineral waters in different systems, as well as the best methods of using them in certain pathological conditions of the system, must, as with all other medicines, be learned from observation. Now, as physicians but rarely have an opportunity of observing the use of mineral waters for a sufficient length of time and in a sufficient variety of cases, and as but little has been written by those who have observed their effects, it ought not to be supposed that the medical public generally would be greatly enlightened on this subject.

I have said that the opinions generally prevailing in enlightened circles relative to the curative powers of our principal mineral fountains, being founded on experience, may, in the main, be correct. I would not be understood, however, as advising a reliance upon such "popular fame." Information of this kind is sufficient to awaken attention and incite inquiry, but certainly should not be implicitly relied upon in individual cases. At best, it is generally "hearsay" opinion, made up, ordinarily, from partial and empirical sources; or, quite as likely, from the prejudiced accounts which are brought by visitors from the different watering-places, and which are *sweepingly* favorable, or prejudicial, as they may chance to have been benefited or worsted, and that without reference to the specific action of the agent, or that clear understanding of the pathology of the case, which would serve as a safe guide in its application to others. Every physician knows how prone persons are to err in the use of medicines, from the supposed resemblance of cases. Often am I pained to see persons persevering in the use of a mineral water to their evident prejudice, and for no better reason than that Mr. or Mrs. Such-a-one was cured of a disease supposed to be similar; or, by the general recommendation of some medical man who sent them to the "mountains" with a "*carte blanche*" to use "*some of the mineral waters.*" Occasionally it has become my painful duty to advise patients to retrace their melancholy steps homeward, without using any of the waters, because none were adapted to their case.

Mineral waters are not a *panacea;* they act, like all other medicines, by producing certain *effects* upon the animal economy, and upon principles capable of being clearly defined. It follows, that there are various diseases and states of the system to which they are not only *not adapted*, but in which they would be eminently injurious.

Some years since, I was requested to visit a highly-respectable gentleman, who had just arrived at the

White Sulphur with his family, from one of our distant cities. He was in wretched health, and sought my advice as to the applicability of the water to his case. On examination, I felt astonished that any medical man of intelligence should have recommended such a case to mineral waters for relief. I advised the gentleman to retrace his steps homeward, and put himself under medical treatment, as he had no time to lose. Accordingly, the ensuing morning he recommenced his journey of seven hundred miles to reach his home. Medicine did for him what mineral waters were not calculated to do, and I have since heard of his entire recovery. This gentleman informed me that he had been influenced to undertake the distant and, to him, painful journey, by a physician who had never before prescribed for his case, and who candidly stated to him that he knew but little of the mineral waters of Virginia, but he had heard of many cures from their use, and therefore advised that he should hasten to give them a trial. Influenced by this vague opinion, the unfortunate invalid had dragged himself and his family seven hundred miles, under the vain hope of finding a remedy, which the physician should, in such a case, have found in his own office. Now, a little more knowledge of the nature of mineral waters, and a more commendable caution in advising their use, would have prevented the heavy sacrifice this gentleman incurred. Nor is this by any means an isolated instance; my case-book furnishes many others equally strong, that have come under my observation in the course of my practice.

Medical Efficacy, etc.—Mineral waters are exceedingly valuable as medicinal agents, are applicable to a large circle of cases, and will, unquestionably, cure many which the ordinary remedies of the shops will not. Nevertheless, it should always be borne in mind that they are not a *catholicon;* that they are not to be used for every disease; and that, to be prescribed successfully, they must, like all other medicines, be prescribed *with reference to the nature and pathology of the case.* Nor is this caution ordinarily more necessary in using the various medicines of the shops than in using the more potent mineral waters.

Some there are, I know, who profess to be unbelievers in the medicinal activity of mineral waters, and who, without denying the benefit that is often derived from visiting such fountains, attribute the whole to travel, change of air, exercise, relaxation from business, etc. Now, I freely admit that these are often important agents in the cure of a large class of cases; but, from long experience at a popular watering-place, and the numerous cures I have seen effected from the water itself, totally disconnected with any of the adjuncts alluded to, it would be quite as easy to convince me that *bark* is not tonic, that *jalap* does not purge, or that *mercury* will not salivate, as that mineral waters may not be an active and potent means of curing disease, entirely independent of the valuable adjuvants that have been alluded to.

The advocates of the non-efficacy of mineral waters, *per se*, would scarcely persist in this opinion, after seeing the large amount of active medical material obtained by evaporation from some of our more active waters; the *White Sulphur*, for instance, which yields more than one hundred and fifty grains to the gallon, and which, upon analysis, is found to consist of *iodine, sulphur,* the various combinations of *soda, magnesia,* and other active ingredients. Would it not be absurd to believe that so large an amount of these efficient medical substances as is usually taken into the stomach, by those who drink mineral waters in which they abound, could fail to exert a *positive influence* upon the economy? My own experience for many years, in the use of such waters, enables me to bear the most unequivocal testimony as to the *direct* and positive influence of many of them upon the human body. In the language of the celebrated Patissier, I can unhesitatingly say that, "in the general, mineral waters revive the languishing circulation, give a new direction to the vital energies, re-establish the perspiratory action of the skin, bring back to their physiological type the vitiated or suppressed secretions, provoke salutary evacuations either by urine, or stool, or by transpiration; they bring about in the animal economy an intimate transmutation—a *profound change;* they saturate the sick body. How many sick persons, abandoned by their physicians, have found health at mineral springs! How many individuals, exhausted by violent disease, have recovered, by a journey to mineral waters, their tone, mobility, and energy, to restore which, attempts in other ways might have been made with less certitude of success!" And hence most cordially do I adopt the sentiments of the distinguished Dr. Armstrong, who, in speaking of the medicinal efficacy of mineral waters, says, "*I dare pledge my word, that, if they be only fully and fairly tried, they will be found among the most powerful agents which have ever been brought to the relief of human maladies.*"

Modus Operandi, etc.—Various attempts have been made to account for the peculiar effects of mineral waters upon the system. They seem to act, in the first place, as a simple *hygienic* agent. Secondly, they act, in accordance with their constituent ingredients, specifically on the animal economy. Mineral waters exert their more important influences upon the human body upon a different principle from many of the articles of the materia medica; they are evidently *absorbed*, enter into the circulation, and change the consistence as well as the composition of the fluids; they course through the system, and apply the medical materials which they hold in solution, in the most minute form of subdivision that can be conceived of, to the diseased surfaces and tissues; they reach and search the most minute ramifications of the capillaries, and remove the morbid condition of those vessels, which are so commonly the primary seats of disease. It is thus that they relieve chronic disordered action, and impart natural energy and elasticity to vessels that have been distended either by inflammation or congestion; while they communicate an energy to the muscular fibre and to the animal tissues generally, which is not witnessed from the administration of ordinary remedies.

Many of the articles of the materia medica seem to act by sympathy and counter-irritation, and to cure one organ of the body by irritating another; thus calomel, by irritating the stomach and duodenum, is made to act efficiently upon the liver, to which organ it has a strong specific tendency. Not so, however, with mineral waters: *they never cure one organ by irritating another.* I can with confidence assert, that I *have never seen mineral waters successfully used in any case in which they kept up a considerable irritation upon any of the organs of the body.*

Both physicians and patients are far too much in the habit of looking to the *immediate* and *sensible operations* of mineral waters, and of judging of their efficacy from such effects. In most cases, it is serviceable for such agents to open the bowels gently; and in some, it is best for them to purge actively. Occasionally, advantage is derived from promoting an increased flow of urine or perspiration; but, as a general rule, the greatest good is derived from the *absorption* of the water, resulting in that "profound change" spoken of by Patissier, or, in other words, the *alterative* action of the remedy. It should always be borne in mind that this *profound change*—this *alterative effect*—is incompatible with constant or active action of the water upon any of the emunctories. This, unquestionably, is true as relates to the *White Sulphur water*, and I believe it to be so with all alterative waters.

So well convinced am I, that the *alterative action* is the real curative action effected by *sulphur waters*, in nine cases out of ten where any serious disease exists, that, ordinarily, I am not solicitous to obtain much daily increase of evacuation from any of the emunctories. On the contrary, I often find great advantage from the administration of some appropriate means to *prevent* the too free action of the water, especially on the bowels and kidneys. As a general rule, it is far better that such waters should *lie quietly upon the system*, without manifesting much excitement upon any of the organs, and producing, at most, but a small increase in the quantity of the ordinary healthy evacuations.

The *quality* or kind of evacuations produced by mineral waters is a matter of far more importance, and, when strong sulphur waters are used, never fails to evidence the existence and the extent to which alterative action is going on in the system; and to this, persons using such waters should always pay a careful attention.

I have said that the best effects of mineral waters are their *alterative* or *changing* effects; and that, in the administration of the White Sulphur, I do not, ordinarily, desire to provoke much increase of the natural evacuations. I do not wish, however, to be understood, by this general declaration, as laying down an absolute rule of practice to govern all cases, or to apply in reference to all waters. The administration of mineral waters, like the administration of every other remedy, should be governed in reference to the particular character and demands of each case; and in such discriminating practice it will sometimes be found best to use them in a manner to produce active operations for a short time. I have, indeed, generally found, that those who are actively purged by mineral waters, if they have strength to bear it, will be best satisfied with the remedy *at the time*, and, in fact, are apt to feel better *at the time*, than those upon whom the water is exerting but little or no purgative effect. It may be laid down as a general fact, in the use of all *alterative* waters, subject to but few exceptions, that those on whose bowels they act freely will feel best *while at the Springs;* while those who are but little purged will feel best after they have *left the Springs*, and will, ordinarily, enjoy the most permanent advantage. The reason of this is obvious: in the first case, the active purgation throws off the gross humors of the body, and the patient feels promptly relieved; in the other case, the remedy lies upon the system, is absorbed, and gradually produces its changing influences,—bringing the various secretory functions into a healthy condition,—unloading and cleansing the machinery of the economy,—silently putting its *works* to rights, and giving them their natural and healthy motion. All this requires time for its accomplishment; and hence, we often hear persons say, "I was no better while at the Springs, but I began to mend soon after I left, and have continued better since." Declarations of this kind we constantly hear by persons who have previously visited alterative springs; and they verify the correctness of my proposition.

Length of Time to be used, etc.—To acute diseases, mineral waters are not adapted; for all such they are too exciting, too prone to increase the activity of the circulation, and to stimulate the general system. It is in *chronic* diseases only that they are found so eminently serviceable. By chronic diseases I mean those slow diseases of the system uniformly attended either with *simple excitement*, chronic *inflammation*, or chronic *congestion* of the blood-vessels. To be permanently beneficial in diseases of this description, the use of mineral waters, like the disease for which they are taken, should be "chronic." I mean that an instantaneous cure should not be expected; but that the remedy should be persisted in, and the cure gradually brought about. Sulphur waters, especially, may be easily brought into disrepute by short and imperfect trials of them. To prove effectual, "they should for the most part be continued daily, in sufficient quantity, until the disease gives way, or until their inefficacy has been fairly proved by an unremitted perseverance. In some cases of *ophthalmia*, of *rheumatism*, and *slight cutaneous affections*, I have known them to effect a cure in two or three weeks, while in other cases, apparently similar in all respects, twice, thrice, or even four times that period has elapsed before the cure had been accomplished; and what is here affirmed of these external affections, is still more strongly applicable to internal diseases, which are seldom speedily overcome by these waters, how completely soever they may yield at last. In illustration of this point, as to internal diseases, it may be mentioned that I have seen both chronic inflammation of the liver, and chronic inflammation of the rectum, where no benefit was produced for three or four weeks, and yet a *continuation* of the waters for six or eight weeks longer has effaced every vestige of the morbid indications for which they were prescribed." (*Armstrong on Sulphur Waters.*)

There is no greater folly, in the use of mineral waters, than that of laying down *a definite period of time for which they should be used*, without reference to their effects upon the system. Like all other medicines, mineral waters should be used, discontinued, or modified in their use, with a strict regard to their operations upon the body, and to their good or bad effects upon the disease. Whenever prescribed, their operations should be watched with the same care with which we watch the effects of any other medicine; and they should be persevered in, or temporarily or permanently discontinued, or controlled in their action by some appropriate adjuvant, according to the indications presented in each case.

It will occur to every reflecting mind, that the expectation of being cured, or even essentially benefited, in an *obstinate chronic disease*, from a few days' use of any mineral water, is altogether unreasonable. Nevertheless, I have often seen persons at watering-places despairing of the efficacy of the remedy, simply because it had not produced an obvious and appreciable benefit in five or six days. A sort of *stereotyped* opinion indeed prevails with numerous visitors to such places, that the water should not in any case be used longer than two weeks. I scarcely need say that this is a most erroneous opinion, and often interposes between the patient and his recovery. It is true that some, who hold the unwarrantable opinion alluded to, perseveringly endeavor to drink as much in the "two weeks" as they should do in six; but this only serves in a common way to make them abandon it four or five days before their prescribed time, by absolutely disqualifying the system for its reception at all.

I can say, as the result of many years' observation, that the *White Sulphur*, which is one of the strongest sulphur waters in the world, rarely produces its full *alterative* effects within two weeks, under its most judicious administration, and under favorable circumstances for its use; and that three, four, five, and even eight weeks often elapse before it has displayed its full remedial powers in obstinate cases. And such will be found to be the case with all alterative waters.

General Remarks on the Administration, etc.—Mineral waters are all *stimulants* in a greater or less degree, and some have attributed much of their virtue to this property. Such an opinion, however, is clearly erroneous. I have already remarked that such waters are rarely serviceable when they keep up any considerable irritation of an organ. I now remark that any considerable excitement of the general organism is equally prejudicial; indeed, I have often been embarrassed, and sometimes thwarted in the successful use of mineral waters, from the prevalence of this quality.

The amount of excitement resulting from the use of such waters depends upon the nature of their constituent principles; upon the quantity taken, the manner of taking it, and the excitability of each individual's constitution. If it be a water abounding in sulphuretted hydrogen gas, the most essential difference exists in taking it *with* or *without its gas;* that is, in taking it fresh at the spring, or after its gas has flown off. In the use of the *Sulphur Waters*, with or without their peculiar gas, the most marked difference exists in their stimulating quality, and it is greatly advantageous in many cases, particularly in very excitable persons, to have the gas expelled in part, or in whole, before using them.

Some mineral waters, by varying the method of their administration, or by the interposition of appropriate adjuvants, are capable of extensive and valuable modified actions and effects upon the human body. The White Sulphur is susceptible of as many varied, different, and modified actions upon the system generally, and upon its particular organs, by varying the methods of using it, as is *mercury*, or *antimony*, or any of our leading therapeutical agents. For instance, it can be so used as to *stimulate* distressingly; or, without any *appreciable stimulating effect*. It can be so given as almost invariably to *purge actively;* or, without lessening the quantity producing such effect, but merely by changing the time and manner of taking it, it can be so given as to exert little or no cathartic operation. It may be directed to, or restrained from, the *kidneys*, or skin; and what, in a general way, is far more important, it can be so used as to *lie quietly* on the system, producing no excessive action upon any of the organs, and, with a quiet but sure progress, go on breaking up the obstructions in the glandular organs and removing the impediments to the proper discharge of their functions: equalizing the circulation, removing chronic inflammations, and generally restoring the energies of the system.

CHAPTER II.

MINERAL WATERS IN GENERAL.

CONTINUED.

Resemblance of some Mineral Waters to Mercury—Errors and Abuse of Mineral Waters, etc.—Changing from Spring to Spring—Dress—Diet, Exercise—Best Time for Using—Length of Time to be Used, etc.

Resemblance to Mercury, etc.—Between the action of *mercury*, and the more powerful of the sulphur waters, on the organic system, the most striking similarity exists. Dr. Armstrong long since remarked the resemblance between mercury and the sulphur waters of Europe, and confidently expressed the opinion that the latter are equally powerful with the former, in their action upon the secretory organs; and with this very important difference, that while the long-continued use of mercury, in chronic disease, generally breaks up the strength, that of the sulphur waters generally renovates the whole system. Mercury has heretofore, by common consent, been regarded as the most powerful alterative we possess. I am not prepared to dispute this high claim of the medicine, but this much I will assert, as a matter of professional experience, that sulphur water, in my hands, has proved an *alterative* quite as certain in its effects as mercury, though somewhat slower in its operations. Not only so, I believe it to be far better adapted than mercury to a large circle of cases, in which glandular obstructions and chronic inflammations are to be subdued. If the claims of the two remedies for preference were otherwise nearly equal, the great advantage on the score of safety from the sulphur water would give it an immense preference over its rival. Numerous cases present themselves, however, in which they are used in conjunction to great advantage. Where this becomes necessary, I have, as a general rule of practice, found it best not to continue the mercury longer than six or eight days; nor is it often necessary to use it continually during that period.

The effects of the *White Sulphur water* upon the human body resemble mercury in several respects. Not to mention others, its resemblance is strikingly manifest from the fact of its producing *salivation** under certain peculiar circumstances. Another marked similarity may be mentioned, especially as it has a direct bearing upon the proper method of its administration: I allude to the existence of a phlogistic diathesis in individuals with whom either remedy is used. When the system resists the specific action of mercury, it is a certain test that the inflammatory diathesis prevails to a considerable extent, and this is the cause of the resistance; for lessen the inflammatory diathesis by proper evacuations, and the specific action of the mercury will be readily induced. The system often offers the same resistance to the successful use of this water, which is evidently occasioned by the excess of the inflammatory diathesis, inasmuch as when the inflammatory disposition is abated by the lancet, purgatives, etc., the water promptly produces its wonted good effects. In the administration of this particular water, it is of the utmost consequence to keep this practical fact constantly in view, and, by proper treatment, to keep down both general and local excitement.

Notwithstanding mineral waters are so well adapted to the cure of chronic diseases, it should not be ex-

---

* Dr. Salsbury, the resident physician at Avon Springs, has witnessed similar effects from the *Avon* water.

pected that they will be uniformly successful; for it must be remembered that such diseases are only remediable when unconnected with alterations of organic tissue, which is their ultimate and mortal product. Nor is it reasonable to expect that any plan of treatment will succeed in all cases of chronic disease, unconnected with alteration of tissue; and I have accordingly found the methods recommended at times ineffectual, even when they were tried under circumstances which simply indicated disorder of the function, without any concomitant sign of disorganization.

Errors and Abuse of Mineral Waters, etc.—I have before alluded to some of the abuses of mineral waters by those who resort to them for relief; this subject, I conceive, may be still further pursued with profit to my readers. To one familiar with the many errors and mistakes committed in the use of mineral waters in this country, it will not seem wonderful that numbers return from visiting our most celebrated watering-places without having received any essential benefit, but be rather a matter of surprise that so large an amount of good is achieved. The precautions in the use of such waters, deemed indispensable in France, Germany, and England, are greatly neglected here. There, the advice of a competent physician, who is well acquainted with the nature and peculiarities of the water, is thought so important, that persons rarely enter upon their use without such *advice*, and, at some places, are actually not *permitted* to do so. If similar precautions were more commonly adopted by visitors at our various watering-places, a far larger amount of good would be secured to the afflicted, much injury prevented, and the character of the several waters better established and preserved. It is a subject of daily and painful observation, at all our principal watering-places, to witness numerous individuals using mineral waters that are not adapted to their cases; and still more common is it to see those, to whose cases they are adapted, using them so improperly as entirely to prevent the good they would accomplish under a proper administration. Professor Mütter, of Philadelphia, makes the following judicious remarks when speaking of the use and abuse of mineral waters in this country: "Like every other remedy of any efficacy, mineral waters are liable to abuse, and it is really astonishing that such glaring errors should be daily committed, not only by the patients, but often by the *physicians* who recommend their employment. It is by no means an uncommon occurrence (and those who have visited the springs of our country will bear me out in the statement I am about to make) for an individual to arrive, furnished with a '*carte blanche*,' from a physician who has probably little or no knowledge of the active properties of the agent he recommends, to use the water as he may *see fit*, or with merely a charge to '*use it with caution*.' Others are sent without any direction whatever, in the hope that the water *may suit* their condition, and come trusting in Providence alone. Others, again, arrive with written instructions to drink so many glasses of the water *per diem*, whether it agrees with them or not. Many patients do not take the advice of a physician at all, but, relying on the representations of those who have derived benefit, imagine that they, too, will be cured, although, in all probability, from the nature of their disease, the water may be the most prejudicial to which they could resort. Used in this careless and dangerous manner, is it to be wondered at that so many individuals leave the springs either not at all benefited, or in a worse condition than when they arrived?

"The regulations which are thought necessary, and which are adopted in most European countries, especially France and Germany, during the use of a mineral water, are either unknown or neglected in this. There, nearly every spring is supplied with an experienced physician; one familiar with the character of the water, whose duty it is to take charge of the sick as they arrive; here, with but one or two exceptions, those who frequent our watering-places have to rely on *chance* for medical aid. Is this as it should be?"

A vague impression seems to pervade the public mind, that mineral waters, as medicinal agents, are totally unlike all other medicines, and that, in their administration, there is no necessity for observing any cautions, or for adopting extraneous expedients to procure the best effects of the agent employed. This is an error as injurious as it is common, and ought to be corrected in the public mind. Our more potent mineral waters ought indeed to be regularly incorporated into our *materia medica*, their several qualities properly defined, and the medical mind thus instructed to regard them, not only as valuable therapeutical agents, *per se*, but as agents capable of extensive and valuable modifications in their application to disease. A *pathological practice* should be established in relation to them, not less strict than in relation to the ordinary remedies of the shops, and the best means of influencing their sanative operations on the system understood.

The physician who desires to throw his patient under the *alterative* influence of mercury, is not so discouraged as to abandon the remedy, if it chance at first to run off by the bowels, and thus thwart his object; but, either by changing the method of using his medicine, or by uniting with it some soothing astringent, he ultimately effects the important object in view. Neither should the patient be discouraged in the use of a mineral water because it occasionally manifests a vagrant and improper effect; for facilities can be commanded to control its operations, as readily as we can control the improper operations of mercury. Such facilities may generally be found, either in an *increase* or *diminution* of the quantity taken,—an alteration of the *periods* at which it has been taken,—or in the manner of using it (where gases prevail), in relation to its *gaseous* or *ungaseous* form. Occasionally medical adjuvants are found necessary, and then I have been in the habit of using those most simple, and those which least derange the animal economy.

As a general rule, I have found mineral waters most serviceable in those cases in which the stomach and general system tolerated them readily; yet such toleration depends so much upon the proper *preparation of the system*, and the manner of using the water, that the patient should by no means infer that it is unsuited to his case simply because it has manifested some improper operation in the commencement. For, as before intimated, it will often happen, that by changing the method of using the water, or by the administration of some appropriate medicine, the difficulty will be removed, and the water afterwards act most pleasantly and profitably upon the system.

Liability to Mistake in Reference to Sulphur Waters.—While on this subject, it is not inopportune, I conceive, to allude to a popular and common error in reference to the *quality* of sulphur waters in general,—an error into which the intelligent as well as the ignorant are prone to fall: I allude to the very common mistake of forming a judgment as to the strength and value of a sulphur water merely from its *taste* and *smell*. Most persons who have not carefully investigated the subject are ready to believe that they have discovered a valuable sulphur fountain when they have found a water abounding in sulphuretted gas. This, as a general thing, would be a mistake, and, as it is a mistake that might lead to a profitless use of such waters by invalids, it seems proper that attention should be distinctly called to it.

I have elsewhere* sufficiently contested the idea that

---

* Chapter on the "Relative Influence of the Gaseous and Solid Contents of the White Sulphur Water."

sulphuretted hydrogen gas ought to be regarded as an efficient medicinal agent, except so far as its nervine and stimulant qualities give it such claims. I do not now propose to go over the arguments for the correctness of this opinion,—they are sufficiently set forth in the chapter alluded to,—but merely to enter up this *caveat* for the benefit of sulphur water drinkers,—that *the mere fact of water being strongly impregnated with sulphuretted gas* is not, *of itself*, a sufficient evidence that it is a valuable remedial agent.

We often see waters abounding in this gas, and, to the taste and smell, very much resembling the best of our standard waters, and hence imagined by many to be identical in quality and equal in strength to them, but which, upon trial, are ascertained to have but little medicinal value, and are found, by analysis, essentially *without body*, with little efficiency in their medicinal salts; or, with a *combination of saline matters* not well adapted to give them medicinal virtue.

Neither does the color nor abundance of deposits made by such waters, as they flow from their source, do more than afford a problematical evidence of their value.

First. Because it is to the *quality* of the saline matters, rather than to their abundance, that we are to look for medicinal efficacy; and,

Second. Because the color of the natural deposits of all sulphur waters, unmixed with foreign bodies, as I have elsewhere said, is always essentially the same, being invariably white or opaque-white; the various shades of blue, gray, red, black, etc., being occasioned by the influence of light and shade, or being chemical changes, occasioned by their coming in contact with foreign bodies.

The color of the deposits of such waters, it will be seen, then, cannot to any degree indicate their quality or value. A large amount of deposit of saline matters, yielded by any mineral water, is strong *presumptive* evidence of its strength, but is not conclusive evidence of its medicinal value, in the absence of a knowledge of the peculiar *quality* and *combination* of such saline matters. Hence we should not hastily judge of the value of a mineral water by the *color* of its deposits, nor even by the *large amount* of its deposits, but by their *quality*, and the proportions in which they are relatively combined in the water, forming a compound suited to the great mission of modifying and healing disease.

Springs are occasionally found that abound, either largely or sparsely, in sulphuretted gas, and that contain but little saline salts; and yet such springs are often valuable for particular forms or types of disease, and are rendered so from the quality and fortunate combination of their salts. On the other hand, waters may abound largely in saline matters, and some of these saline matters be valuable, too, as single agents, yet the entire compound which they form may not be well adapted for sanatory and medicinal influences.

CHANGING FROM SPRING TO SPRING.

A very common error, in the use of Mineral Waters, is the belief that the patient should often change from one water to another, and that no one should be used longer than some given number of days, and this without any reference to its effects upon the system. This absurd notion leads many persons to fly from spring to spring, performing in a few weeks or days the circuit of the whole "*spring region*," and without remaining long enough at any one to receive permanent benefit. Now, if the position heretofore laid down be correct, that "mineral waters, like all other medicines, cure disease by exerting *effects* upon the animal economy," the impropriety will be obvious to all of rapidly hastening from one fountain to another, without tarrying long enough at any to receive those *effects* upon the body which are necessary to a cure. Such a water-drinker acts like the "maid of all work," always busy, but accomplishing nothing.

What would be thought of the physician who, having decided that his patient must undergo the influence of alterative action upon his system, and having put him upon a course of mercury to accomplish the object, should, just before this drug would have accomplished the end, discontinue its use, and put him upon iodine; and, just as this was about to alterate the system, abandon it and substitute sarsaparilla; and thus, from one drug to another, running through the whole routine of alterative remedies, without giving any sufficient time to effect the object? This would surely be an absurd method of practice; and yet it would not be more absurd than the course we often see pursued by visitors at mineral springs,—who literally waste their whole time in going from fountain to fountain, and thus debar themselves of all permanent good, by spending their time rather *among the springs* than at any one of them. The state of mind, which leads invalids thus improperly to act, is often induced from the random opinions or injudicious advice of their fellow-sufferers, whom they meet with at the various watering-places. One will tell another that they have seen or heard of some person that was cured at once, at this, that, or the other spring. Among the Virginia springs, for instance, you will be assured by one that the "White" is the place; by another, that the "Salt" is better suited to your case; a third informs you that you would do better at the "Blue;" while others will tell you there is nothing like the "Red," the "Sweet," the "Warm," the "Hot." Thus are the minds of persons frequently perplexed, until they come to the conclusion to "make the rounds," and try them all for a day or two. In this way the hapless invalid is often led to fritter away the whole time he remains in the mountains, without deriving permanent advantage from "*all the springs,*" when, very probably, the time he had fruitlessly spent at them all would have been sufficient to cure him at *any one of them.*

Let it be distinctly understood that these remarks are meant for the serious invalid only. Persons who visit the springs for amusement or pleasure, or those who go merely as a relaxation from business, and require only the tone which travel and mountain air can give, may, with great propriety, go from spring to spring, and spend their time just where they are the happiest. But for the invalid *who has something for the waters to do*, it is not so; he should first wisely determine which of the springs is best calculated to cure his disease, and, having settled this important question, should persevere in the use of that particular water, carefully watching its effects, and "not be carried about by every wind of doctrine," but continue the use of the agent thus wisely selected, either until its inapplicability has been proven, or until it produces the specific effects which he desires. This being accomplished, there may be, and often is, a necessity for visiting other springs.*

DRESS.

Delicate persons, visiting the mountains or colder latitudes for health, should be particularly cautious on the subject of dress. It is rather more easy to dress with the ever-varying fashions, than to dress appropriately for *all the weather* that happens in mountainous regions generally, during the watering seasons. The weather, in such situations, is often so variable and uncertain as to make it a good general rule for the invalid to dress without reference to any particular state of it, but always warm and comfortable, with (in most cases) but little change from his dress in the spring season before he reaches the mountains.

---

* See chap. iii., on "Prescribing Mineral Waters."

Some invalids will be benefited by constantly wearing soft flannel next the skin, not only because it keeps up a more uniform temperature than linen, but also because of the gentle excitement it occasions on the surface of the body. The best summer dress, however, which I have ever seen worn next the body,—and always a valuable *accompaniment of flannel*, winter and summer,—is *woven silk*. I am led to believe, from experience, that silk, worn next the skin, is the very best protection we can command against the influence of cold. In *rheumatism* and *neuralgia*, a covering of woven silk is a valuable *remedy;* and for all delicate persons, and for those peculiarly susceptible to colds, it is a most invaluable shield to the body. The superiority of silk over every other covering is probably owing to its peculiarity as a non-conductor of electricity; but whether this be so or not is left to the astute medical philosopher to determine; it is sufficient for me to know the fact of its superior efficacy, without stopping to account for it.

Since the above paragraph was first written, I have had twenty-five years' additional observation of the use of silk as a covering for delicate and susceptible persons; and the result is, that I am more than ever convinced of its great superiority. Indeed, such persons, while in our variable climate, and under the influence of sulphur waters, that increase the susceptibility of the system, cannot, by any other dress, so effectually secure themselves against the encroachment of cold, as by the use of *silk sacks* worn next the skin. Nor ought this precaution to be neglected by such, especially as the existence of a cold always renders the use of the waters less efficacious, and sometimes positively injurious, for the time it may continue.

DIET, EXERCISE, ETC.

Diet and exercise, during the use of mineral water, are of too much importance to be passed over without notice. It is to be regretted that so little, as relates to diet, is placed within the power of the invalid at our watering-places generally. Usually there is but one general system of living at all such places, and this invariably a system very ill adapted to the invalid.

Persons using mineral water may ordinarily indulge, in moderation, in that diet which they found to agree best with them at home. Imprudences as to the kind of food, or of excess in its quantity, should be as carefully avoided by the invalid while using such water, as when under treatment by other medical means. This, however, is by no means commonly the case.

Mineral waters generally remove acidity from the stomach, and sharpen both the appetite and the digestion; hence it is often really difficult for the invalid to restrain himself at table, and we might be astonished to see the quantity and quality of food he sometimes consumes. Dyspeptics, as might be expected, suffer most from impropriety in diet; indeed, I am persuaded that more than half the good these waters would otherwise achieve, in such cases, is prevented by impropriety in diet. But the evil of over- and improper feeding, although most manifest in dyspeptics, is by no means confined to such. Upon the subject of diet, Dr. Bell has well observed, that "slow and laborious digestion, heartburn, disordered kidneys, discoloration of the skin, and some affections of the liver, often the effects of excessive eating and drinking alone, are not to be readily cured by visiting mineral springs, and keeping up the same kind of living." If they (and the remark applies to all invalids) be sincerely desirous of gaining health, they will most successfully do so by simplifying their regimen, and abstaining from all those appliances to force appetite and tickle the taste, which they had formerly used in the shape of ardent spirits, wine, and malt liquors, fried meats, pastry, and unripe fruits. In fine, we may sum up in a few words, by repeating, after the great father of medicine, that *all excesses are dangerous;* a maxim every one must have fully tested.

Eating much in the evening, sitting up late, prolonged and immoderate dancing, remaining too long in the cool air of the evening, are often the cause of many unpleasant complaints, which might have been easily prevented.

The passions are to be kept in check by avoiding every exciting cause, either of the boisterous or melancholy kind. A giddy chase after pleasure and luxurious indulgence are scarcely more reprehensible than an indolent and secluded life. The kind and amount of exercise to be indulged in by the patient must, of course, be regulated by the nature of his disease and the attendant circumstances; walking, riding on horseback or in a carriage, may be selected, as one or the other may be best adapted to the physical ability and to the inclinations of the patient; but, in some form or other, all whose strength will admit of it should take regular exercise in good weather.

PERIODS FOR THE USE OF MINERAL WATERS.

The best time for using mineral waters is in the morning before breakfast, when the stomach is empty and the absorbent vessels are most active. They may generally be used to advantage an hour or two before dinner, and before going to bed at night.

In many cases it is best that the whole that is taken in the course of the day be divided into two parts, and taken, either in the morning before breakfast, and a short time before dinner, or in the morning, and a short time before going to bed at night.

Advantage is not often secured by such waters taken before supper; and often such use of them—except a very moderate use—is prejudicial, from their proneness, when thus taken, to run off by the kidneys.

As a general rule, mineral waters, and especially

alterative waters, have their best effects when taken *before breakfast*, and before *going to bed at night*. There are some, however, who do not bear them well at night; and attention should always be paid to this circumstance.

Such waters should not be used immediately before or after a meal; nor should glass after glass ordinarily be taken in *quick succession*. By such imprudent use the stomach is overtasked, and unpleasant consequences result, such as *eructations*, giddiness, an unpleasant sense of fullness, and sometimes permanent injury of the stomach with *atonic dyspepsia*.

LENGTH OF TIME TO USE MINERAL WATERS.

The length of time invalids should continue the use of mineral waters depends entirely upon the *nature of the case* for which they are used, the *manner of using*, and the *susceptibilities of the system*. Some believe that they will exert all their sanative influences in a given number of days; and then should be discontinued. The use of such waters should not be limited to a given number of days without careful regard to *effects*. Some cases will be thrown as fully under their curative influences in two weeks as others will be in four, or even eight; and yet they may be equally well adapted to each case. In every case of their administration, respect should rather be had to the *effects* they are producing than to the time they have been administered.

They never cure disease until they have first produced *effects* upon the system,—EFFECTS which can always be distinguished by the experienced observer, during the progress of their operation, with the same certainty with which he can distinguish the effects of any of the articles of the materia medica.

It often happens that invalids use mineral waters that are well adapted to their cases, and use them assiduously for several weeks, without deriving a particle of permanent benefit; and this in consequence of so improperly using them, both *in time* and *quantities*, as to force the water out of the system by the emunctories, *without touching the case*,—without being permitted to tarry long enough to produce those *salutary effects* which must precede a cure. This is especially true in reference to waters that cure disease mainly through their *alterative* influences.

The range of time within which the full effects of mineral waters may be expected is from *two* to *eight* weeks, according to the *nature of the case, a proper use of the remedy, and the general susceptibility of the party using them.*

Sulphur waters, that benefit mainly through their *alterative* powers, require a somewhat longer use to produce their full effect than do the *saline, acidulous,* or *ferruginous* waters. I have very rarely *seen* the full *alterative effects* of the White Sulphur attained within *two weeks;* and have generally found that from *three* to *six* weeks' persevering use of it was necessary to insure its full effects in confirmed and obstinate cases.

CHAPTER III.

USE OF MEDICINES AND DIFFERENT MINERAL WATERS.

Prescribing Mineral Waters.

THE judicious administration of mild and appropriate medicines, in connection with the use of mineral waters, with the object of facilitating their operations upon the system, is often a matter of primary importance.

All writers who treat of mineral waters as medicinal agents urge upon invalids the propriety of obtaining experienced medical advice before commencing their use, and allude to the occasional necessity of using medicines in connection with them in obstinate cases. But the circumstances under which medicines should be used, and the primary necessity of the practice in particular cases, have not always been as fully insisted on as the merits of such practice demand. This, we suppose, has been owing rather to the positions occupied by the various authors on mineral waters, than to any want on their part of a proper appreciation of the subject. A portion of such authors, although learned and scientific men, and highly distinguished in their profession, have not, nevertheless, had a large actual experience in the treatment of disease at mineral fountains and with mineral waters. Hence the teachings of such have, very properly, been designed to show the value and adaptation of such agents as *independent* remedies, rather than as important adjuvants in particular cases; consequently they have treated of them in a somewhat isolated sense, and as they would have treated of any single article of the materia medica. The few who have written upon the subject, whose residence at mineral fountains has afforded enlarged opportunities for investigating the peculiar effects of the waters in individual and diversified cases, may, to some extent, have been restrained by motives of delicacy from enlarging upon this subject as fully as they should have done. Such authors, being settled as practitioners at the fountains of which they write, may not unnaturally have felt, that for them to urge upon the invalid visitor the necessity of medical advice and assistance, however important they might esteem it, and with however much of candor and disinterestedness they might do so, would possibly subject them to invidious reflections by the illiberal, or even from the discreet stranger, who, not fully appreciating the importance of the subject, might misapprehend their well-meant motives.

Many persons are disposed to regard mineral waters, in their curative powers, as a *panacea*, and, like the much-extolled catholicons of the day, unaided by other appliances, and in despite of scientific directions and all the rules of art, adapted to cure all manner of diseases. I need scarcely say that such opinions, when entertained, are very erroneous, and that the judgment which regards them as important remedies in *nature's materia medica*, having, indeed, a wide and valuable scope of operation, but, like all other remedies, necessarily demanding various modifications and cautions in their use, would be far more correct and reliable.

Many consecutive years of experience, in the administration of mineral waters, have given me great confidence in their employment; indeed, I yield to no one in admiration of their happy adaptation for many ills to which flesh is heir. As *independent* remedies, totally disconnected with all other medicinal aid, they are often fully sufficient to attain the sanative end desired. So, too, we occasionally find a single article of the *materia medica*, unaided by other articles, capable of producing every beneficial effect that the case demands. Doubtless, like results occasionally take place from the employment of the various panaceas or catholicons of the age. But where we meet with one case in which a single article of the *materia medica*, or an artificial panacea, unaided by all other means, satisfactorily fulfills all indications of treatment in chronic disease, and results in effecting a cure, we meet with perhaps ten cases in which adjunctive remedies should be employed. Be this as it may, however, in reference to the remedies just alluded to, we know it to be true of alterative mineral waters, not only as to the *certainty*, but especially as to the *celerity*, with which they effect cures in obstinate cases. This view of the subject is not only consonant with reason, but also with the general theories and teachings of the profession.

There is an opposite view of the subject, however, which alleges that any medical agent, adapted to the case, is sufficient of itself for the case, and should therefore stand unassisted by any other means. This theory, it will be perceived, leads necessarily into empiricism, and to the discarding of all science and discrimination in the use of remedies; and, consequently, ignores the value of all knowledge and experience in the profession.

Now, I admit that if the selected agent be so fully and entirely adapted as really to fill every indication in the case, then the proposition I am combating is true,—and under such circumstances every judicious physician would say, *let it alone*. But such full and complete adaptations are but occasionally found to exist, either in medicines or mineral waters; and, in the use of the latter, even under ordinary happy *adaptations*, we often find a state of things that primarily existed, or has been superadded, that must be remedied by appropriate medicines, or the water, so far from proving beneficial, will act injuriously. Besides, admitting the mineral water to be never so well adapted to the case in which it is being used, its slow progress in resolving congestions and in overcoming diseased action may, in many cases, be greatly hastened by judicious adjuvants, skillfully and timeously administered.

In obstinate cases in which it is desirable to procure the specific operations of a mineral water upon any organ, much time, to say the least, is saved by uniting with the water, for a few days, some *adjuvant* that *specifically determines to such organ*. By such a procedure, the water may be *invited* to the organ, and establish its action upon it much sooner than it would without such aid.

In diseases of the abdominal viscera generally, the patient may often economize a week or more of the time which otherwise it would be necessary for him to use the water, by the proper introduction of some medical adjunct to the end that has been intimated.

The proportion of invalids, especially of such as are suffering with biliary derangements, that will derive increased benefit from the employment of mild alterative cathartics, to precede or accompany the use of alterative mineral waters, is as *ten to one at least;* and, in nine cases out of ten, the subject of biliary derangements will *economize a week or ten days*, in the necessary use of such waters, by the occasional use of medicines.

The general rule, which may with safety be laid down for the guidance of those about to use mineral waters, is to have their stomach and bowels well cleansed of fæcal and mucous collections, and to bring down, as near as may be, the circulation to a natural standard.

A medical rule, in attempting the cure of disease, is to subdue inordinate and evident disturbance of the system before we administer medicines with a view to their peculiar effect. Thus, when the stomach and bowels are highly irritable, or inflamed, we decline administering purgatives; when there is acute pain in the head, with high fever, we withhold opium and other remedies of what are termed the class of anodynes; when the liver is acutely inflamed, we are wary in giving anti-bilious medicines, so called. Violent and regularly recurring chills do not justify the use of the barks, if the interval be marked by symptoms of high action of the blood-vessel system generally, or of great determination to the head, liver, or stomach. All these several states of violent disease are to be mitigated before we enter upon specific remedies. Without preliminary treatment in the cases supposed, purgatives would, so far from carrying off matters oppressive to the stomach and bowels, and promoting secretions from their inner surfaces, only serve still further to irritate and inflame these parts; opiates would increase the pain in the head and restlessness, and even cause delirium; bark would convert the remittent into more of a continual fever, and increase the distress of the stomach, and exasperate the prior existing pain in the liver.

From these and other analogous facts, we learn the important truth,—overlooked by the public generally, and sneered at by impudent quacks,—that the operations and remedial effects of any one medicine, or combination of medicines, *are purely relative, and depend on the state of the animal economy at the time.* These views should be carefully borne in mind, as well in the administration of mineral waters as of the ordinary remedies of the apothecary's shop.

I desire not to be misunderstood, however, as expressing the opinion that medicines are always necessary in ordinary cases submitted to the use of mineral waters.

When the powers of the water are sufficient to answer, with tolerable certainty and celerity, the sanative indications, it is safe, and generally proper, to withhold medical means altogether; or, if occasionally any should be demanded, to employ such only as are mild and suasive in their character.

PRESCRIBING MINERAL WATERS.

The medical adviser at popular watering-places has, necessarily, very delicate and responsible duties devolved upon him. To some extent he must be the recipient, in a professional point of view, of the confidence of the invalid stranger who has left a distant home, to seek at medicinal fountains the best remedy for the maladies of which he hopes to be relieved. This confidence, while it is agreeable to the honorable mind, is not without onerous responsibility.

A sufficient knowledge of our various mineral springs, to enable the medical adviser to judge correctly of their specific character and adaptations, unfolds at once to him a wide field for the exercise of skill and judgment, in selecting for his patient the one best adapted to the nature and wants of his case.

In the Virginia Spring region, for instance, we are surrounded by a perfect galaxy of mineral fountains, of almost every variety and adaptation. We have the *Sulphur* waters, in their various modifications; we have the *Chalybeates*, simple and compound, in great variety; the *Saline*, in several varieties; the *Aluminous*, or acidulated aluminous chalybeates, in three or four varieties; and *thermal* waters of every temperature, from 62° to 106°. All these fountains of healing, with their varied modified influences (for each one differs in some essential particulars from all the others), should be regarded as so many different articles in nature's *materia medica*, each possessing adaptations somewhat peculiar to itself, for the different diseases or states of the system. Here, then, is a wide range for the medical adviser, and his tact and success, in advising most wisely, will necessarily depend upon his acquaintance with the peculiar qualities and specific effects of all these different agents.

Again, such an adviser, to be most useful to his patients, must be careful not to be influenced by his *loco personæ*, or to regard the particular fountain over whose medical direction he presides, as a catholicon, and adapted better than any other to all sorts and conditions of cases. A medical adviser, at a mineral fountain, could not well fall into a greater error, or more clearly evidence a want of wise discrimination, than in finding his remedy, in all cases, in the particular agent which he immediately directs; for, in the nature of things, such universal preference would often be misplaced. Standing in the delicate relation which such an adviser holds to the invalid public, he must regard the various mineral agencies around him somewhat in the same light in which he regards the various medicines of the apothecary's shop, and should wisely and freely choose among them for the use and benefit of his patients. Any other course would be empirical, —hazardous to the best interest of the unfortunate invalid, and utterly unworthy of his confidence.

Under such proper and discriminating advice, the patient will often, perhaps in a majority of cases, be led in the course of his cure to the use of several of the different fountains. The same water, however potent it may be, is not always, nor even generally, sufficient to meet all the indications that exist in the case, and, unaided, to produce a perfect cure. There is nothing more common than the certainty with which a particular water accomplishes particular results upon the animal economy, while it fails to accomplish other results that will be readily achieved by other and dissimilar waters. For instance, while some waters are well adapted to produce *alterative* effects upon the secretory organs, and, by their general emulging and changing influences, to bring the system into a natural or physiological type,—actions and influences that are primary in their importance, and essential to a cure; this being accomplished, some of the more *tonic and nervine waters* will be found far better adapted to strengthen the animal fibre and to complete the cure.

Potent waters, through the whole catalogue of springs, have each their sphere of usefulness, that must not be overlooked by the discriminating adviser in the treatment of particular cases; and hence they all should be arrayed and labeled, as it were, in nature's great laboratory, and prescribed intelligently, and as their use is indicated in the variety of diseases that are sought to be healed by such agents.

THE BEST PERIOD OF THE YEAR FOR INVALIDS TO VISIT THE SPRINGS.

From the 1*st of June to the middle of July* is preferable to an earlier or later period of the season. There are substantial reasons why invalids should make their visits within the range of time mentioned, and why they should prefer an *early* rather than a *late* period of this range of time.

1st. Because during this period we have, at our watering-places generally, the most delightful weather of the season,—neither too warm nor too cool for exercise in the open air.

2d. Because the crowd of mere pleasure-seekers has not set in up to this period; and hence they are less crowded, and all the facilities and comforts of a quiet home are more easily and certainly obtained.

3d. In the early period of the summer solstice, just after the cold and inclement weather of winter and early spring, and before the sufferer has become enervated by the heat of summer, *chronic disease* more readily yields to the alterative influence of the waters, and, consequently, the invalid is more certainly and speedily placed under their curative powers; and,

4th. Because invalids, whose maladies have been essentially modified or cured in the early part of the summer, have a longer period of favorable weather in which to perpetuate and confirm their amendment and final cure, than those who might receive influences equally beneficial, but obtained at a later period of the summer.

I might allude to other advantages enjoyed by the invalid who makes his visit to mineral waters early in the season; but let it suffice to remark that my long observation as a medical director of such waters has abundantly satisfied me of the decided advantage that attaches to early rather than late visitation by those who are seeking to secure the largest amount of benefit from their use. Hence I earnestly suggest to *invalids* who design visiting mineral waters, not to postpone their visit to a late period of the season, and to *choose an early rather than a late* period of the time I have designated as preferable.

CHAPTER IV.

WEST VIRGINIA AND VIRGINIA SPRINGS.

IN treating of the springs of West Virginia and Virginia, I shall not be guided by their chemical classification, nor strictly by their medicinal importance, but in accordance with their location in the geographical divisions of these States.

The Springs strictly pertaining to what has long been known as the *"Spring Region"* will be first noticed; next, those located in or contiguous to the great *Shenandoah Valley*, formed by the Appalachian chain of mountains on the west, and the Blue Ridge Mountain on the east. Then will follow those found on the eastern slopes of the Blue Ridge and in the plane country stretching towards the ocean, known as *Eastern Virginia*. Lastly, those located in the southwestern counties of the State, commonly known as Southwest Virginia.

The Virginia and West Virginia Springs present great variety in chemical and therapeutic character, comprising various and differently compounded *sulphur* waters; the *chalybeates*, simple and compounded; the *acidulous* or *carbonated;* the *saline;* the *aluminated chalybeates*—with *thermal* waters, varying in temperature from 62 to 106 degrees of Fahrenheit.

Of these Springs, the *sulphurous* waters are found in greater abundance and in greater strength immediately on the western and eastern slopes of the Alleghany Mountains, the strongest being on their western declension. The *simple chalybeates* are found in every great section of both States, but in greatest strength along the course of the great Appalachian range, extending from the northeastern to the southwestern extremities of both of them.

The *acidulous* or *carbonated waters*, as well as the *aluminated chalybeates*, exist in the greatest variety and strength in the central portions of the Great Valley, in the counties of Augusta, Rockbridge, Alleghany, Monroe, and Craig, but are found in several other counties, south and west, along the course of the Alleghany and Blue Ridge Mountains. Waters more or less distinctly belonging to the *saline* class are found in the same range of country.

The most abundant mineral waters in these States, except the simple chalybeate, are the aluminated chalybeates, or *alum waters* as they are commonly called. They are generally found adjacent to faults in the strata, or where the rocks give evidence of derangement from their natural position, and near the junction of *slate* with limestone. They are invariably, I believe, an infiltration through talcose slate which lies a few feet below the surface of the earth. I have examined numerous specimens of these waters, obtained from various neighborhoods, from the head-waters of the Shenandoah River to the extreme eastern border of Tennessee, and have found them to possess the leading chemical characteristics of the springs of this class that have been brought into popular use.

I believe that all the mineral waters in this great range of disturbance are slightly thermal, compared with the temperature of the common springs in their vicinity. But the boundary of the *thermal waters*, commonly so called, is only about fifty miles in length and of narrow dimensions, having the Hot and Warm Springs for its northern, and the Sweet Chalybeate and Sweet Springs for its southern extremes.

ROUTES TO THE PRINCIPAL WEST VIRGINIA AND VIRGINIA SPRINGS.

The results of the war between the Northern and Southern States so materially deranged traveling facilities to many of these Springs as to make the following directions essential to parties at a distance who desire to visit them.

Travelers from the North or East to any of the principal Springs in the mountains of West Virginia or Virginia, to avail themselves most largely of railroad facilities, must necessarily make STAUNTON a point in their journey.

From *Staunton*, the *Rockbridge* and *Bath* Alum, the *Warm*, *Hot*, *Healing*, *White Sulphur*, *Salt*, and *Red Sulphur* Springs, are conveniently reached by railroad, with small amount of staging, and in the order in which they are here set down. The *Sweet* and *Red Sweet* are on the same general route, and are reached by a detour of seventeen miles from the White Sulphur.

The *Yellow*, the *Montgomery White*, the *Alleghany* and *Coiners* Springs, are reached by the traveler *going East* on the Atlantic, Mississippi and Ohio Railroad in the order in which they are here enumerated.

Western travelers to the *White Sulphur*, or other Springs in their region, may reach them most conveniently from *Louisville* or *Cincinnati*, by boat to *Huntington* on the Ohio River, from thence by the *Chesapeake and Ohio Railroad* to the Springs.

CHAPTER V.

WHITE SULPHUR SPRINGS.

Location and General Physical Characteristics—Its Strength uniformly the same—Does not lose its Strength by parting with its Gas—Does not deposit its Salts when Quiescent—Its Gas fatal to Fish—Its Early History—Known to the Indians as a "Medicine Water"—First used by the Whites in 1778—Progress of Improvements, and present Condition—Analyses of Mr. Hayes and Professor Rogers.

THE White Sulphur Springs are located in the county of Greenbrier, West Virginia, on Howard's Creek, and on the immediate confines of the "Great Western Valley," being but six miles west of the Alleghany chain of mountains, which separates the waters that flow into the Chesapeake Bay from those which run into the Gulf of Mexico.

The waters of the spring find their way into Howard's Creek, two hundred yards from their source, which, after flowing five miles, empties into Greenbrier River.

The spring is situated on an elevated and beautifully picturesque valley, hemmed in by mountains on every side. *Kate's Mountain*, celebrated as the theatre of the exploits of a chivalrous heroine in the days of Indian troubles, is in full view, and about two miles to the south; to the west, and distant from one to two miles, are the *Greenbrier Mountains;* while the towering *Alleghany*, in all its grandeur, is found six miles to the north and east.

The spring is in the midst of the celebrated "Spring Region," having the "Hot Spring" thirty-five miles to the north; the "Sweet," seventeen miles to the east; the "Salt," and "Red," the one twenty-four, the other forty-one miles, to the south; and the "Blue," twenty-two miles to the west. Its latitude is about 37½° north, and its longitude 3½° west from Washington. Its elevation above tide-water is two thousand feet. It bursts with unusual boldness from rock-lined apertures, and is inclosed by marble casements five feet square and three and a half feet deep. Its *temperature* is 62° of Fahrenheit, and remains uniformly the same during the winter blasts and the summer's heat ; any apparent variation from this temperature will be found, I think, to be owing to the difference in thermometers, as repeated trials with the same instrument proved the temperature to be uniform.

The principal spring yields about thirty gallons per minute; and it is a remarkable fact that this quantity is not perceptibly increased or diminished during the longest spells of wet or dry weather; while other bold springs of the country have failed during the long droughts of summer, this has invariably observed "the even tenor of its way." There is no discoloration of the water during long wet spells, or other evidence that it becomes blended with common water percolating through the earth. The quantity and temperature of this spring being uniform under all circumstances gives a confidence, which experience in its use has verified, of its uniform strength and efficacy. The water is clear and transparent, and deposits copiously, as it flows over a rough and uneven surface, a *white*, and sometimes, under peculiar circumstances, a *red* and *black*, precipitate, composed in part of its saline ingredients. Its *taste* and *smell*, fresh at the spring, are those of all waters strongly impregnated with sulphuretted hydrogen gas. When removed from the spring, and kept in an open vessel for a sufficient length of time for this gas to escape, or when it has been *heated* or frozen for this purpose, it becomes essentially *tasteless* and *inodorous*, and could scarcely be distinguished, either by smell or taste, from common limestone water. Its cathartic activity, however, is rather increased than diminished when thus insipid and inodorous.* It does not lose its transparency by parting with its gas, as many other waters do; nor does it deposit its salts in the slightest degree when quiescent, not even sufficiently to stain a glass vessel in which it may be kept.

The *gas* of this spring is speedily fatal to all animals, when immersed even for a very short time in its waters. Small fish thus circumstanced survive but a few moments, first manifesting entire derangement, with great distress, and uniformly dying in less than three minutes.

The water is uniform in its saline strength; that is, it contains in a given quantity, at all seasons, the same amount of solid contents. Of this fact I am fully satisfied, from repeated tests and examinations of it, under various circumstances, and for many years. It exhibits occasional and slight variations in the amount of its free sulphuretted hydrogen gas. This variation is occasioned mainly, if not entirely, by the condition of the atmosphere at the time, and principally by its electrical condition. Even this variation in the water, however, is more apparent than real, and is often suspected when it does not actually exist.

In the absence of chemical tests, the difference in the water is judged of entirely by *taste* and *smell*, principally by the latter; and some conditions of the atmosphere being more favorable than others for the evolution and diffusion of the gas, the actual relative amount in evolution is often misjudged.

The springs are surrounded with mountain scenery of great beauty, and blessed with a most delightful climate in summer and fall. Independent of the benefit that may be derived from the waters, a better situation for invalids during the summer months can scarcely be imagined. They have the advantage of a salubrious and invigorating air and an agreeable temperature,—cool at morning and evening, the thermome-

---

* See chap. vi., on "The Relative Virtues of the Saline and Gaseous Contents of the White Sulphur Water."

ter ranging at those periods, during the summer, between $50°$ and $60°$, and rarely attaining a greater height than $85°$ at any time of the day,—with an elasticity in the atmosphere that prevents the heat from being at any time oppressive, and enabling the invalid to take exercise in the open air during the day without fatigue.

There is but little in the early history of this watering-place especially worthy of preservation.

Tradition says that the charming valley in which it is situated was once a favorite "*hunting-ground*" of the proud *Shawanees*, who then owned and occupied this fair region; and the numerous ancient graves and rude implements of the chase, that are found in various parts of the valley, sufficiently attest the truth of this legend. That a small marsh, originally contiguous to the spring, was once a favorite deer and buffalo "lick," is well known to the oldest white settlers in the country; and it is confidently asserted by some of that venerable class that the spring was known to the Indians as a "*medicine water*," and that since their migration across the Ohio they have occasionally been known to visit it for the relief of rheumatic affections. Whether this legend be truth or fiction, I cannot avouch; authentic history, however, abundantly testifies to the reluctance with which its ancient owners abandoned this lovely valley to the rapacious avarice of the invading white man.

During the year 1774, the proud but ill-fated Shawanees, being overpowered by the encroaching colonists from Eastern Virginia, and having sustained, in October of that year, a signal defeat by the colonial troops, at Point Pleasant, were forced finally to abandon their country, and seek shelter and protection with the main body of their tribe, then living on the waters of the great Scioto; not, however, until, by frequent battles and midnight murders, they had testified their attachment to their ancient hunting-grounds and the graves of their fathers.

The property on which this spring is situated was originally patented to Nathan Carpenter, one of the earliest pioneers of the country, who was subsequently killed by a band of marauding Indians, at a fort at the mouth of Dunlap's Creek, near where the town of Covington now stands.

The precise time at which this spring, now so celebrated among mineral waters, was first used for the cure of disease, cannot be ascertained with absolute certainty. It is believed, however, that a Mrs. Anderson, the wife of one of the oldest settlers, was the first white person who tested its virtues as a medicine.

In 1778, this lady, being afflicted with rheumatism, was borne on a litter, from her residence, ten or fifteen miles, to the spring, where a tent was spread for her protection from the weather; and a *"bathing-tub"* provided, by felling and excavating a huge tree that grew hard by. Here she remained until she entirely recovered, drinking from the fountain, and bathing in the water previously heated in the trough by "hot rocks." It is reasonable to suppose that the fame of this cure spread abroad among the "settlers," and from them into Eastern Virginia, and among the few "spring-going folks," who then annually visited the Sweet Springs, not many miles distant. Accordingly, in 1779, and from that to 1783, there were annually a few visitors here, who spread their tents near the spring, no house having then been erected, and with the rude "trough" for a bathing-tub, and this protection from the weather, are reported to have spent their time most agreeably and profitably. Some of these primitive visitors, "who dwelt in tents," have visited the springs of late years, and, with pleasurable emotions, marked out the spot where their tents stood some sixty years ago, while they recounted with delight the amusements and pleasures they then enjoyed.

In 1784, 1785, and 1786, numerous "log-cabins" were erected, not where any of the present buildings stand, but immediately around the spring,—not one of which, or the materials which composed it, is now remaining.

Mr. Caldwell, until recently the proprietor of the property, came into possession of it in the year 1808, but did not personally undertake its improvement until the summer of 1818. Before this period, the buildings for the accommodation of visitors, although sufficient for the number that then resorted to the place, were exceedingly rude, being altogether small wooden huts. The interest and enterprise of the owner soon led him into a different and more appropriate system of improvement, and from small beginnings he went on, progressing in the rapid ratio of demand, until from the "tent" accommodations in 1779, and the "log-cabins" in 1784, the place now, both in elegance and extent, exhibits the appearance of a neat and flourishing village, affording comfortable and convenient accommodations (including the surrounding hotels) for two thousand persons.*

ANALYSIS.

In the winter of 1842, Mr. Augustus A. Hayes, of Massachusetts, made an analysis of the White Sulphur water, at his laboratory in Roxbury, from a few bottles of water forwarded to him from the spring in the preceding fall. The following is the result of his examinations:—

"Compared with pure water free from air, its specific gravity is 1.00254.

"50,000 grains (about seven pints) of this water

---

* In the spring of 1857, the White Sulphur property was sold to a company of gentlemen residing principally in Virginia, who (in virtue of an act of the Legislature) have associated themselves into a *joint-stock company*, under the name of the "*White Sulphur Springs Company.*" They have erected the largest building in the Southern country.

contained, in solution, 3.633 water grain measures of gaseous matter, or about 1.14 of its volume, consisting of—

Nitrogen gas	1.013
Oxygen gas	.108
Carbonic acid	2.444
Hydro-sulphuric acid	.068
	3.633

"One gallon, or 237 cubic inches, of the water contain 16 739-1000 cubic inches of gas, having the proportion of—

Nitrogen gas	4.680
Oxygen gas	.498
Carbonic acid	11.290
Hydro-sulphuric acid	.271
	16.739

"50,000 grains of this water contain 115 735-1000 grains of saline matter, consisting of—

Sulphate of lime	67.168
Sulphate of magnesia	30.364
Chloride of magnesium	.859
Carbonate of lime	6.060
Organic matter (dried at 212° F.)	3.740
Carbonic acid	4.584
Silicates (silica 1.34, potash .18, soda .66, magnesia, and a trace of oxid. iron)	2.960
	115.735

"Unlike saline sulphuretted waters generally, this water contains a minute proportion of chlorine only, the sulphates of lime and magnesia forming nearly ten-elevenths of the saline matter.

"The alkaline bases are also in very small proportion, and seem to be united to the siliceous earths in combination with a peculiar *organic matter.* The organic matter, in its physical and chemical character, resembles that found in the water of the Red Sulphur Springs, and differs essentially from the organic matter of some thermal waters.

"In ascertaining its weight, it was rendered dry at

the temperature of 212° F. When dry, it is a grayish-white, translucent solid. When recently separated from a fluid containing it, it appears as a thin jelly or mucilage, and gives to a large bulk of fluid a mucous-like appearance, with the property of frothing by agitation. It unites with metallic oxides and forms compounds both soluble and insoluble. In most cases an excess of base renders the compound insoluble. The compound with oxide of silver is soluble in water; with baryta and lime it does not form a precipitate, while magnesia forms with it a hydrous white gelatinous mass. In acids it dissolves; the oxy-acids do not change its composition, while they are diluted and cold; by boiling they produce sulphuric acid from its constituent sulphur, and change its carbon to other forms. In contact with earthy sulphates at a moderate temperature, it produces hydro-sulphuric acid, *and to this source that acid contained in the water may be traced.* This substance does not rapidly attract oxygen from the atmosphere, and from colored compounds, as some other organic compounds do. The proportion of organic matter, like that usually contained in our waters, is in this water very small; until forty-nine-fiftieths of the bulk of a quantity is evaporated, the residual matter does not become colored, and, when the saline residue is dried, it is of a pale yellow.

"The medicinal properties of this water are probably due to the action of this organic substance. The hydro-sulphuric acid, resulting from its natural action, is one of the most active substances within the reach of physicians, *and there are chemical reasons for supposing that, after the water has reached the stomach, similar changes, accompanied by the product of hydro-sulphuric acid, take place.*"*

---

* See chap. vi., on "The Relative Virtues of the Saline and Gaseous Contents of the White Sulphur Water."

Professor William B. Rogers also analyzed this water. The following is the result of his examinations:

Solid matter, procured by evaporation from 100 cubic inches of the water, weighed, after being dried at 212°, 65.54 grains.

Quantity of each solid ingredient in 100 cubic inches, estimated as perfectly free from water:

Sulphate of lime	31.680 grains.
Sulphate of magnesia	8.241 "
Sulphate of soda	4.050 "
Carbonate of lime	1.530 "
Carbonate of magnesia	0.506 "
Chloride of magnesium	0.071 "
Chloride of calcium	0.010 "
Chloride of sodium	0.226 "
Proto-sulphate of iron	0.069 "
Sulphate of alumina	0.012 "
Earthy phosphates	a trace.
Azotized organic matter blended with a large proportion of sulphur, about	5 "
Iodine, combined with sodium or magnesium.	

Volume of each of the gases in a free state, contained in 100 cubic inches:*

Sulphuretted hydrogen	0.66 to 1.30 cubic inches.
Nitrogen	1.88 cubic inches.
Oxygen	0.19 "
Carbonic acid	3.67 "

---

* 100 cubic inches amounts to about 3½ pints.

CHAPTER VI.

THE RELATIVE VIRTUES OF THE SALINE AND GASEOUS CONTENTS OF THE WHITE SULPHUR WATER.

SPECULATION has existed as to the relative efficacy of the different component parts of the White Sulphur water in the cure of disease; and while some have supposed that its *gaseous contents* are essential to its sanative virtues, others, and I think the best-informed observers, attribute its medicinal virtues mainly to its *solid* or *saline contents*. To the latter opinion the able Professor of Natural Philosophy in the University of Virginia, who has carefully examined the water, and other distinguished chemists and physicians, decidedly incline.

It certainly is a question of interest to the valetudinarian, whether he should use this water fresh as it flows from the spring, abounding in all its stimulating gas, or whether he should use it after it has *partially* or *entirely* parted with this gas. To this subject I have devoted particular attention, having instituted, with care, various and diversified experiments, in order to establish something like definite and positive conclusions.

Although the value of this water in what is usually termed its *non-stimulating form*, or, in other words, when deprived of its gas, has long been known to many who are familiar with its use, it was not until the last few years that it was commonly used from choice, after it had been long removed from the spring, or from any cause had parted with its gaseous contents; and an opinion, the correctness of which had never been examined, prevailed in the minds of many, that in losing its gas it lost its strength and efficacy.

Having settled at the "White," as the physician of the place, it became alike my duty and my interest to investigate the character and operations of its waters under every possible form and modification in which they could be presented. In the pursuit of this duty, I resolved to take no opinion upon "trust," but carefully to examine and investigate for myself. A prominent question immediately presented itself for inquiry, involving the relative merits which the *solid* and *gaseous* ingredients of the water possess as remedial agents. It would be tedious, and to many uninteresting, to detail the several steps and multiplied experiments which led me to conclusions upon the subject, satisfactory to my own mind, and upon which I have established certain practical principles in the use of the water, which have enabled me to prescribe it, especially for *nervous* and *excitable patients*, with far greater success than heretofore. It is sufficient for my purpose at present to state that, while I freely admit that the *gas*, which abounds in the water, is an active *nervine stimulant*, and therefore may be a most potent agent in some cases, we are, nevertheless, to look mainly to the *solid contents* of the water for its *alterative* power, as well as for its activity manifested through the emunctories of the body.

Whether the efficacy of the solid contents be owing to the specific character of any one, or to all of the *thirteen different salts* of which it is composed, and which exist in the water in the most minute form of subdivision, and in this condition enter the circulation, and course through the whole system, applying themselves to the diseased tissues; or whether its efficacy, to some extent, depends upon the *evolution* of sulphuretted hydrogen gas, *after the water has reached the stomach*, is a matter of curious inquiry.

The distinguished chemist, Mr. Hayes, of Roxbury, after having bestowed much pains in analyzing the water, and in studying its peculiar character, comes to the following conclusions as to the source of its medicinal power. After describing, at considerable length, a certain matter which he found to abound in it, and which he terms *"organic matter,"* in the course of which he says, "it differs essentially from the organic matter of some thermal waters," he proceeds to say: "In contact with earthy sulphates, at a moderate temperature, it produces hydro-sulphuric acid, *and to this source that acid contained in the water may be traced.* This substance does not rapidly attract oxygen from the atmosphere, and from colored compounds, as some other organic compounds do; *the medicinal properties of this water are probably due to the action of this organic substance.* The hydro-sulphuric acid, resulting from its natural action, is one of the most active substances within the reach of physicians. *There are chemical reasons for supposing that, after the water has reached the stomach, similar changes, accompanied by the production of hydro-sulphuric acid, take place."**

Before Mr. Hayes had communicated the above opinion, growing out of his chemical examinations, I had again and again been much interested with certain phenomena which I have termed the *secondary formation* of gas in the White Sulphur water. Instances had frequently been reported to me of the water having been put into bottles after it had *lost its gas entirely*, being void both of taste and smell, and yet, after these bottles were kept for some days in a warm situation, and then opened, the water appeared equally strong of the hydro-sulphuric acid, as it is found to be, fresh at the fountain.

In a shipment of this water to *Calcutta*, some years since, the "Transporting Company" had the water bottled in Boston, from barrels that had been filled at the spring six months before. The water, although *tasteless and inodorous,* when put into the bottles at

---

* See Hayes's Analysis, chap. v.

Boston, was found, on its arrival at Calcutta, so strongly impregnated with the hydro-sulphuric acid as to render it necessary, under the direction of an intelligent gentleman of Boston (who had witnessed this secondary formation of gas before), to uncork the bottles for some time before using, that the excess of gas might escape.

I had, also, known that in the process of *thawing* sulphur water, which had been previously frozen, sulphuretted hydrogen gas is evolved; for although the ice has neither the taste nor smell of sulphur, a strong smell of sulphuretted hydrogen is manifest as the ice is returning to water.

I had often observed that individuals who drank the water entirely *stale*, and void alike of *taste* and *smell*, were as liable to have eructations of sulphuretted hydrogen as those who drank it fresh at the fountain. These, and other facts connected with the peculiar operations and effects of the water when used in its ungaseous form,—operations and effects which it is not necessary here to refer to, but all going to prove the *secondary* formation of gas under certain circumstances, —had, in my investigations of this water, interested me exceedingly; and, consequently, I was not a little pleased that Mr. Hayes's chemical examinations so fully sustained the opinions I had been led to entertain from my personal observation.

This opinion of Mr. Hayes, in connection with the numerous proofs derived from analogy and observation, of the *secondary* formation of sulphuretted hydrogen gas in the water, would seem to be calculated to harmonize the opinion advanced by me of the *equal efficacy* of the water when deprived of its gas, with the sentiment entertained by some, that the hydrogen gas is essential to its sanative operations.

The phenomenon of a *secondary formation* of sulphuretted hydrogen gas in mineral waters has not, that I am aware of, been noticed before; it certainly has not been in relation to the White Sulphur, and we hope that medical gentlemen, generally, who may have occasion to use such waters, will direct attention to this singular fact. For myself I promise still further to investigate the subject, and may, at some subsequent period, lay the results of my investigations before the medical public.

My investigations of the relative virtues of the gaseous and saline contents of this water have satisfied me that the physician, in making up his judgment as to the best method of administering it in particular cases, may always properly moot the propriety of using it *fresh* as it flows from the spring,—*deprived of its gas*,—or with *modified quantities*. He should bear in mind that there are cases in which it is preferable that the water should be used *stale*, and that, by depriving it in *whole* or in *part* of its gas, he can graduate that amount of stimulus to the system, which it may demand, and this, in most cases, without lessening the *actively operative* or *alterative* effects of the water.

For some patients, the White Sulphur, as it flows from the spring, is too *stimulating*, and hence, before the *non-stimulating* method of using it was introduced, many such patients left the spring, either without giving the water a trial, or actually rendered worse by its stimulating influence. This class of persons can now use the water, *when deprived of its gas*, not only with impunity, but often with the happiest results. Numerous cures, effected by its use in the last thirty years, have been in that class of patients by whom the water, *fresh at the Spring*, could not have been used without injury.

In cases of nervous persons, and especially in those whose *brain* is prone to undue excitement, I have often found it necessary, either by *freezing* or *heating* the water, to throw off its gas completely, before it could be tolerated by the system; and some of the happiest results I have ever witnessed from the use of the water have been achieved by it after being thus *prepared*.

My object in prescribing White Sulphur has been to pursue a discriminating or *pathological* practice. I regard it as an active and potent *medicine*, and believe that, like all such medicines, it should be used with a wise reference to the nature of the case and the state of the system. I *must not be understood as advancing the opinion, that this water is always to be preferred after the escape of its gas.* I entertain no such opinion; on the contrary, for a large class of visitors, I think it preferable that they should avail themselves of the use of the water either at, or recently removed from, the fountain, and as it naturally abounds in its gases. There are other cases in which the exciting influence of the gas can only be borne *in a more limited degree*, and, for such, I permit its *partial escape* before using it; while in a numerous class of cases (and especially on first commencing the use of the water) I esteem it indispensable to its quick and beneficial operation, that its *uncombined gas*, which gives *taste* and *smell*, should have escaped.

In recommending the White Sulphur, then, to the use of the invalid, I esteem it quite as necessary to investigate the manner of using, as relates to its *fresh* or *stale* quality, as in reference to its dose, or the times of administering it; and for neither would I lay down positive and absolute rules in advance; for each case must, in the nature of things, give rules for its own government.*

---

* It is now more than thirty years since the author first called public attention to the importance, indeed, the absolute necessity, in many cases, of the invalid's using this water in its *ungaseous* or *least* stimulating form.

Like all innovations upon old opinions and customs, it met with its hasty objectors, at first, but actual experience was not long in establishing the soundness and value of the recommendation, and now I have the gratification to know that it is regarded by all well-informed persons as a *fixed principle* in the use of the water, that, to be used safely and most beneficially, in very many cases, it must be used with strict reference to its *fresh or stale* quality; or, in other words, to its *stimulating* or *non-stimulating* effects.

The great value of this water, as a therapeutical agent, to a large class of persons who visit the fountain, is a fact alike unquestioned and unquestionable. That in its natural condition, as it flows from the bosom of the earth, it is happily adapted to numerous cases of disease, is a truth established by upwards of eighty years' experience and fully sustained by the numerous cures that are constantly occurring. The value of the water, then, fresh as it flows from the spring, and abounding in its gas, is a truth, so far as I know, that is *unassailed*, and which, I believe, is *unassailable*. Nevertheless, that there are many cases in which the gas is not beneficial, *in the amount* in which it exists in the fresh water, is a fact which my experience enables me to assert with the utmost confidence. That the water, in such cases, therefore, is better without its gas than with it, follows as effect follows cause. But I do not teach that the water, *per se*, and without reference to cases, should always be preferred without its gas. I base not my practice upon any such narrow and exclusive views; nor do I deny the value of the agency of the gas in appropriate cases.

I, then, regard the *solid contents* of the White Sulphur water, either in its direct or indirect influences, as the *main* agency in its medicinal efficacy. Whether the *efficacy* of the salts of the water be owing to their absorption into the system as such, or whether it depends upon the *secondary formation* of hydro-sulphuric acid gas in the stomach, or whether it ought to be ascribed to the combination of these different agencies, I leave for others more fond of speculation to decide. I have, heretofore, been satisfied with the *knowledge* of the efficacy of the solid contents, without much theorizing to explain the *why* and *wherefore*.

But, it may be asked, if the gas does good in the state of a *secondary formation* in the stomach, would not a larger quantity, taken with the fresh water, do more good? I reply, that this by no means follows in that class of cases for which I specially advise the ungaseous water; for my only objection to the fresh water, in such cases, is, that it has *too much gas*. Admitting that the gas may exert an influence, I allege that in nervous and excitable cases the quantity is not only better adapted to the system, but that any given quantity, under a *secondary formation*, excites the system less, from its gradual formation in the stomach, than if suddenly received in volume into that viscus.

Nor do I, because I recommend the ungaseous water in *particular cases*, repudiate and disallow all medicinal agency of the gas, as a general principle. Not at all. I simply contend that, *for the treatment of certain cases*, there is *more of the stimulating gas* in the fresh water than such cases can bear with advantage, and that its excessive excitation in such cases would be prejudicial instead of beneficial.

But do I find it necessary to guard the amount of gas for every water-drinker? or in effect to erect a bed of *Procrustes* and oblige every one to conform to its length? By no means. A. arrives at the springs, not much debilitated by disease, and with a firm nervous and muscular system; there is no excessive excitability in his case, and neither his cerebral, nervous, nor vascular system is particularly prone to be affected by stimulants or exciting medicines. I advise him to use the water *as it flows from the fountain*, and if he should, contrary to expectation, find that it stimulates him unpleasantly, to set it by for a short time before using.

B. calls for advice as to the manner of using the water; his *temperament*, and the state of his cerebral, nervous, and vascular system, are the opposite of A.'s; his physical energies have been prostrated by disease; his nerves are *unstrung*, and, like his brain, prone to be painfully affected by stimulants or exciting medicines. He is advised to use the water after it has, either *partially* or *entirely*, parted with its gas; that is, after it has been set by for *twelve* or *eighteen hours*, as the delicacy and excitability of his system demand.

In cases of inflammation of the *parenchyma* of the brain, and in other highly excitable conditions of the cerebral or nervous system, I have the water more carefully prepared, either by heating or freezing it.

In graduating the amount of stimulus, or, if the gaseous theorist please, the amount of medical material, to the wants of the system,—in other words, *varying the prescription to suit the case*,—am I departing from a scientific and approved system of practice? What would be thought of the science of a medical man who invariably used either the same medicine, or the same dose of any medicine, without regard to the peculiarities or constitution of his patients? Just what ought to be thought of any one who would direct so potent an agent as White Sulphur water to be used *alike* in every variety of constitution and disease.

A popular error, in relation to mineral waters, is that they exert a sort of mysterious influence on the system; and that, as nature has elaborated them in the bowels of the earth, they are, therefore, formed in the best possible manner for the cure of disease. This opinion is not more reasonable than it would be to suppose that nature has formed *antimony* in the best possible form, for the cure of disease, although we know that in this form, under the administration of the celebrated Basil Valentine, it slew all the *monks* in his cloister.

Like all other remedial agents, potent mineral waters produce certain *effects* upon the animal economy, and these *effects* will be beneficial or injurious, as the remedy is properly or improperly employed. For instance, C., who is nervous, delicate, and excitable, and is affected with functional derangement of the organs, requires to receive, for a certain time, the influence of a mineral water, which, while it acts as an aperient upon his bowels, enters his circulation, courses through his system, and *alterates* his deranged organs; being, at the same time, so bland and unstimulating in its general effects, as not to arouse any one or a series of organs into undue excitement and rebellion against the common good. Such a remedy is found in the *stale* and *ungaseous* White Sulphur water.

D. requires the very same effects to be exerted upon his diseased organs,—but he is of very different temperament and constitution. His brain and nerves are prone to no unnatural excitement, and he is unaffected with the thousand physical sensibilities to which C. is subject. D. may take the White Sulphur water with impunity and advantage, in any manner most agreeable to him. In his case its exciting gas constitutes no objection to its use. The good effects of the water, so differently used by C. and D., will be the same, *because the difference in their cases makes the difference in the use of the remedy.*

CHAPTER VII.

GENERAL DIRECTIONS FOR THE USE OF THE WHITE SULPHUR WATER.

Directions meant to be General, not Specific—Must not generally look to the Sensible Operations of the Water for its Best Effects—Moderate or Small Quantities Generally Preferable—Necessary Preparations of the System for the Use of the Water—Sensible Medicinal Effects of the Water—Effects on the Pulse—Synopsis of Rules to be Observed—Use of Baths.

MUCH that might have been said under this head has been anticipated in the chapter on "Mineral Waters in General."

It is scarcely necessary to remark, after all that has heretofore been said of the necessity of using MINERAL WATERS with *strict reference to the nature of the disease in which they are employed*, that it is not designed that the directions herein given shall be considered sufficient to guide in the use of the White Sulphur in all cases, or in any difficult and important case, to the exclusion of the more minute and specific directions which such case may demand. It is my intention rather to indicate the *general rules* which ordinarily must be observed in its administration, than to lay down definite directions which shall apply to all cases.

Every one who is familiar with the various types of disease, and with the peculiarities and radical difference in different constitutions and temperaments, modifying and influencing diseased action, will at once see the impossibility of laying down any *absolute* rule, for the use of a potent mineral water, that should be strictly adhered to in all cases. Each case, to a certain extent, must, with this, as with all other medicinal agents, indicate the proper dose, and the proper manner of administration.

As has been already remarked, it is very common to attribute the beneficial effects of mineral waters to their immediate *sensible* and *obvious* effects upon the human body. I have shown this opinion to be erroneous; that, so far from its being true that such waters uniformly manifest their beneficial effects by their *active operations*, such operations frequently delay, or entirely prevent, the good which they otherwise would have accomplished through the medium of their *alterative* effects.

Those who desire to obtain the *alterative* operations of the water must, as a *general rule*, take it in small quantities, and continue its use for such length of time as will be sufficient, in common Spring parlance, to "saturate the system." Patients thus using the water are apt, however, to become restless and dissatisfied for the first few days; so much so, that it is often difficult to reconcile them to this manner of administration; because, say they, "it is doing me no good;" they wish to see such tokens of activity as are given by prompt and vigorous purgation. In a general way, it is preferable that the water act sufficiently on the bowels, even when given in reference to its *alterative effects*, to obviate the necessity of giving any other medicine for that purpose; but it is often better to use some mild purgative from the shops, to effect this object for the first few days, than that the quantity of water should be greatly increased.

I desire, especially, to call the attention of physicians, and of the intelligent public generally, to this *distinctive alterative quality of the water*. In this, more than anything else, it differs from other mineral waters. Many other waters are found to possess valuable *alterative* power, and with an equal or greater cathartic or diuretic action, but none have yet been shown to be so *certainly, promptly,* and *powerfully alterative* upon the human system.

Some of my unprofessional readers may desire to know the precise meaning that is attached to the term ALTERATIVE, in a medical sense. This term simply means to *alter* or *change;* that is, to alter or change the chemical composition of the blood, the secretions of the glands, and the various secretory organs and surfaces, the removal of obstructions from the glands or minute vessels which occur in congestions, irritations, and inflammations; thus restoring the blood and the general organism to their natural condition and to the performance of their natural functions.

I claim that the water has these effects by being absorbed, or, in other words, entering into the great circuit of the circulation, and thus exercising the specific or peculiar action of its constituents in promoting the various secretory and excretory processes, and thereby restoring the diseased system to a physiological condition.

Such effects and changes, wrought in the sick body, are obviously an *alteration*, and the remedy that produces them is an *alterative*.

This is but a part of a medicinal alterative; but it conveys a sufficient idea of its nature.

The opinion is as common as it is erroneous, among those who visit mineral waters, that they are to be benefited in proportion to the quantity they drink. Persons in health, or not debilitated by disease, do sometimes indulge in enormously large and long-continued potations of such waters, with apparent impunity; but it by no means follows that those whose stomachs are enervated by disease, and whose general health is much enfeebled, can indulge the habit with equal safety. In such stomachs the effects of inordinate distention are always painful and injurious, while the sudden diminution of the temperature, from large quantities of cold fluid suddenly thrown into the system, can scarcely fail to prove injurious.

I sometimes meet with another class of visitors, who err just as much on the opposite extreme; they arrive at the springs, and place themselves under the government of a *recipe* for the use of the water, drawn up, most commonly, by some distant medical adviser, who has never himself had an opportunity of observing its effects; and such not unfrequently take this *aqua medicinalis* in literally *homœopathic doses;*—in quantities altogether insufficient to produce any sanative effect.

PREPARATION FOR THE USE OF THE WATER.

Some preparation of the system, preceding the use of the water, is often, though not always, necessary for its safe and advantageous administration. Most persons, after the excitement usual to the travel in visiting the springs, will be profited by taking some gentle purgative, and by the use of a light and cooling diet for a day or two before the water is freely used. Those in feeble health should commence the water with caution, and generally in its *least stimulating form,*—that is, after it has remained in an open vessel until its gas has escaped. If, with these precautions, it fail to exert its desired effects, or produce unpleasant symptoms, the medical adviser, to whom it would be necessary to resort in such an emergency, would, of course, prescribe according to circumstances; nor can any general rule be given as respects the treatment that would be necessary in such a case,—one patient often requiring treatment essentially different from another.

Invalids, however, ought not to despair of the use of the water, and of its adaptation to their cases, simply because it may, at first, or even in the progress of its use, display some vagrant and improper action upon the system. *Errors in its action, if they may so be termed, generally arise from errors in its use*, and may generally be prevented by a change in the method of administration, or by some medical assistants, so that the water may be safely continued.

SENSIBLE EFFECTS OF THE WATER ON THE SYSTEM.

The sensible medicinal effects of the water are prominently displayed in its action upon the *bowels*, *liver*, *kidneys*, and *skin*, and, when drunk fresh at the fountain, by a lively *stimulant* effect upon the system in general, and upon the *brain* in particular.

Proper quantities, taken in the morning before breakfast, will often exert some *cathartic* effect in the course of the day. The *liver* is, in most instances, brought under its influence from a few days' perseverance in the use of it, as will be manifest from the character of the excretions. Its action upon the *kidneys* is readily induced, and we occasionally see it exerting, at the same time, both a diuretic and a cathartic operation. Very commonly the exhalant vessels of the skin are stimulated to increased *perspiration;* but its full effects upon the surface, manifested not only by increased, but *sulphurous perspiration*, do not occur until it has been freely used for several weeks, nor until the secretory system generally has been brought under its influence.

In reference to its *cathartic* effects, I remark, that while as a general rule it gently opens the bowels, and in some cases purges freely, we meet with occasional cases in which its effects are distinctly constipative from *the first*. In other cases I have known it to purge gently for the first few days and afterwards to produce constipation.

As the system is brought under the influence of the water, the appetite and the ability to digest food are sensibly augmented. The spirits become buoyant and cheerful, with increased desire for social company and amusements.

Exercise, previously irksome, is now enjoyed without fatigue, and so great is the change in the whole man, that the patient often expresses his appreciation of it by declaring that he is "a new man,"—and so he is, in reference to his physical and social feelings.

EFFECTS ON THE PULSE.

The effect of the water upon the *pulse* ought to be distinctly noted, inasmuch as its action upon the circulatory system affords one of the best indications of its adaptation, or inadaptation, to the case.

As a general rule it will be found that, after the water has been properly used for a sufficient time to affect the circulation, by those to whose cases it is well adapted, and the frequency of whose pulse is much above the natural standard, the *pulse will be reduced in frequency and in force.* This reduction of the pulse is not the consequence of any *direct sedative* action of the water on the heart and arteries, but is the sanative result of its alterative and calming influences upon the general economy; and especially from its agency in stimulating glandular secretions, emulging the emunctories, removing offensive débris that oppress the circulatory organs and functions, thus giving a clear and unembarrassed course to the great circuit of the fluids through the system, as well the chyle and lymph as the venous and arterial blood.

A common consequence from the proper administration of the water, in cases to which it is well suited, is an essential modification of the circulation both in frequency and force; so much so, indeed, that I am never surprised to find the pulse, whose beat has been from 90 to 120 in the minute, reduced to 75 or 80, and, in many cases, quite down to the natural standard of the individual, whatever that may have been; while the volume of blood in the artery is increased, as well as the softness and mildness of its flow.

Experience has so clearly taught me to rely upon the reduction of the frequency and force of the pulse, as indicative of the value of the water to the patient, that I habitually look to such effects as among the most distinct indications to persevere in its use.

On the contrary, if the effects of the water be to increase the number of pulsations, or in any considerable degree to render the circulation more irritable, my inferences are unfavorable to its use; and if this state of things cannot be readily changed by a different administration of the water, its discontinuance is advised, for *it never proves beneficial when it perseveringly excites the frequency of the circulation.* There may be a condition of things in the case that would not justify a hasty discontinuance of the water, merely because of its proneness to stimulate, in a slight degree, the heart and arteries; but the propriety of continuing its use, in any such case, can only be safely judged of by the well-informed and discriminating medical mind.

SYNOPSIS OF FACTS ILLUSTRATING THE MEDICINAL CHARACTER OF THE WATER, ETC.

The following facts, intended to illustrate the peculiar medicinal character and influences of the White Sulphur water, as well as the best manner of using it in ordinary cases, have been alluded to in other parts of this volume; nevertheless (although it may involve a repetition), it is thought best to group them under one general head, for the greater convenience of the reader.

Severally, and collectively, they are positions of great importance to the invalid, and long experience enables me to regard them in the light of APHORISMS, or fixed facts.

1. The water is always more *stimulant*, and generally *less purgative*, when taken fresh at the spring and abounding in its gas.

2. The *alterative*, or changing, effects of the water are by far its most valuable effects, and are those which, more than all others, give to it its distinctive and effective character.

3. If the water produces *active purgative* or *diuretic* effects, its *alterative action* is correspondingly delayed.

4. In obstinate and important cases, the invalid should never consider that he has given the water a fair trial, or that he has obtained its full curative effects, until he has experienced its general *alterative influences*, and maintained them upon the system for some time, and *this entirely irrespective of the time he may have used the water*.

5. As it is uniformly true that the water is seldom permanently serviceable, when it acts as an *irritant* upon any portion of the body, it follows that its use should not be persevered in when, for any considerable time, it continues thus to act. It may, however, almost invariably be made to act kindly and soothingly, by a modification of the manner of using it, or by such gentle medicinal appliances as the peculiarity of the case may demand.

6. From an improper use of the water, or from failure to use a timely dose of medicine, to bring the system into a proper condition to receive it, it occasionally disagrees with persons (to whose constitution and case it is well adapted), until the errors, whatever they may be, have been corrected.

7. An active and long-continued *diuretic effect* is generally useless, and frequently hurtful, and hence, when in much excess, should be arrested. This may be effected *with the utmost certainty* by a modification in the *quantity*, or *periods of using the water, and by gentle medical means that divert from the kidneys and determine to the liver and skin*.

8. As to the amount of water to be used in the course of the day, or as to the number of days it should be used, it is impossible to lay down a *definite rule to apply in all cases*. So much depends upon the nature of the case, and the peculiarities of the constitution of the patient, that no *fixed rule* in these particulars can be laid down as applicable to all cases, and an attempt to do so would be an act of empiricism more apt to mislead than to edify.

USE OF BATHS.

A most valuable aid in the use of this water is the *tepid, warm,* or *hot* sulphur bath. I cannot here enter into particular directions for the use of such baths. I just observe that they may be made an important auxiliary in a large circle of cases, if timely and otherwise properly employed.

Hot sulphur bathing, indeed *hot bathing* of any kind, is a remedy potent and positive in its influences;—capable of effecting much good when judiciously employed, or corresponding evil when improperly used. Like potent mineral waters, it is often used empirically and improperly, and hence becomes a curse when it should have been a blessing. It is a remedy essentially revolutionary in its character,—never negative, but always producing positive results upon the economy, for good or for evil.

The condition of the system indicates with sufficient clearness, to the experienced observer, the time for commencing, and the temperature of the bath. In most cases, the *bathing-point* is as clearly indicated under a course of sulphur waters as the blistering- or bleeding-point is in inflammations, and the value of the bath is much dependent upon such timely employment. When the water has well opened the bowels,—has found its way into the general circulation, softening the skin and calming the irritation of the arterial system,—the *baths* may be looked to with confidence in their efficacy.

Hot baths ought never to be taken during the existence of febrile excitement. They should be used on an empty stomach, and, as a general rule, before the decline of the day, and their temperature always carefully regulated to suit the nature of the case and the state of the system.

Persons intending to *bathe in warm or hot sulphur*

waters should, previously to doing so, be intelligently instructed under a proper knowledge of their case, as to the precise *temperature* of the bath, and the *length of time* to remain in it. Neglect or disregard of proper instructions, the relying upon chance, or the mere dictum of ignorance upon this subject, has often been the cause, within my knowledge, of aggravation of symptoms, and, in many instances, of serious consequences. I state, therefore, for the benefit of bathers in sulphur waters, that such baths, to be *used safely and efficaciously*, must be used with careful reference to their *temperature;* the *state of the system when employed;* and the *length of time* the bather remains in them.

CHAPTER VIII.

DISEASES IN WHICH THE WHITE SULPHUR MAY, OR MAY NOT, BE USEFULLY PRESCRIBED.

Dyspepsia — Gastralgia — Water-Brash — Chronic Gastro-Enteritis—Diseases of the Liver—Jaundice—Enlargement of the Spleen—Chronic Irritation of the Bowels—Costiveness—Piles—Diseases of the Urinary Organs—Chronic Inflammation of the Kidneys—Diabetes—Female Diseases: Amenorrhœa, Dysmenorrhœa, Chlorosis, Leucorrhœa—Chronic Affections of the Brain—Nervous Diseases—Paralysis—Some Forms of Chronic Diseases of the Chest, or Breast Complaints (to be avoided in Pulmonary Consumption)—Bronchitis—Chronic Diseases of the Skin, Psoriasis, Lepra, Ill-conditioned Ulcers—Rheumatism and Gout—Dropsies—Scrofula—Mercurial Diseases—Erysipelas—Not to be used in Diseases of the Heart, or in Scirrhus and Cancer—Chalybeate Spring—Effects in Inebriation—Effects upon the Opium-Eaters.

ALL mineral waters, as before remarked, are stimulants to a greater or less degree, and consequently are inapplicable to the treatment of acute or highly inflammatory diseases. This remark is especially true as relates to the White Sulphur, particularly when drunk fresh at the spring, and abounding in its stimulating gas. It is true, as before shown, that when its exciting gas has flown off, it becomes far less stimulating, and may be used with safety and success in cases to which, in its perfectly fresh state, it would be totally unadapted. But even *in its least stimulating form, it is inadmissible for excited or febrile conditions of the system;* and especially to cases of inflammatory action,—at least, until the violence of such action has been subdued by other and appropriate agents.

If the individual, about to submit himself to the use of this water, is suffering from fullness and tension about the head, or pain with a sense of tightness in the chest or side, he should obtain relief from these symptoms before entering upon its use. If his tongue be white or heavily coated, or if he be continuously or periodically feverish, or have that peculiar lassitude, with gastric distress, manifesting recent or acute biliary accumulations, he should avoid its use until, by proper medical treatment, his biliary organs are emulged, and his system prepared for its reception. Much suffering, on the one hand, would be avoided, and a far larger amount of good, on the other, would be achieved, if visitors were perfectly aware of, and carefully mindful of, these facts.

It is an every-day occurrence during the watering season at the "White," for persons to seek medical advice, for the first time, after they have been using the water for days, perhaps for weeks, and it is then sought because of vagrant operations or injurious effects of the water. In most such cases there will be found, upon examination, either the existence of some of the symptoms just mentioned, or evidences of *local inflammation* in some part of the body, sufficient to prevent the constitutional efficacy of the remedy. I am often struck with the control which an apparently inconsiderable local inflammation will exert, in preventing the constitutional effects of mineral waters. To remove such local determinations where they exist, or greatly to lessen their activity, is all-important to secure the constitutional effects of sulphur water.

It is necessary to reflect that mineral waters, like all medicinal substances, are adapted only to certain diseases, and that the more powerfully they act, the greater mischief they are capable of doing if improperly administered; *for, if it be asserted that they are capable of doing good only, without the power of doing harm, we may be satisfied that their qualities are too insignificant to merit notice.*

This consideration indicates the necessity of some caution in the use of waters which possess any sanative powers, and suggests the propriety, in all doubtful cases, of proceeding under the judgment of some professional man who is familiar with the subject, whose judgment may determine how far the water is applicable to each individual case, and in what manner it should be employed to be most efficacious.

A long list of successful cases that have fallen under my care during the third of a century that I have been administering these waters, might perhaps without impropriety be inserted here; but I am induced to omit the insertion, because I am aware with what suspicion medical cases, however well authenticated, are received when they are given to favor any particular practice, or to recommend any particular water. Besides, the insertion of names is objectionable in all private practice, and I consider the reputation of this particular water to be now too well established to require such assistance.

The space I have allotted to this branch of my subject will allow little more than a simple enumeration of the diseases for which this water is beneficially employed. Those who desire more extended information of its effects in the diseases enumerated are referred to my volume upon the "Mineral Waters of the United States and Canada."

DYSPEPSIA.

This common and annoying disease, the especial scourge of the sedentary and the thoughtful, whether existing under the form of irritation of the mucous surface of the stomach—vitiation of the gastric juice—or under the somewhat anomalous characteristic of *Gastralgia*, is treated with much success by a proper course of the White Sulphur water.

The apprehensive and dejected spirit that finds no comfort in the present, and forebodes evil only in the future; the hesitating will that matures no purpose, and desponds even in success; the emaciation of frame and haggardness of visage; the ever-present indurance, and all the imaginary and real ills that torture the hapless dyspeptic, are often made to yield to alterative and invigorating influences that a few weeks' judicious use of the waters has established.

Administered alone, *in every form* of this disease (for under the name *Dyspepsia* we have several *forms of stomach disease* essentially differing from each other, and requiring different modes of treatment), its curative powers may not always be so marked; but in several varieties of the disease, and those indeed which we most often witness, it deserves the very highest praise that can be conferred upon any remedy. In cases of this disease in which the *Liver* is implicated, occasioning slow or unhealthy biliary secretions, a state of things that often exists, the water may be used with especial advantage. To effect *permanent* or lasting cures in dyspepsia, the waters should always be pressed to their complete *alterative* effects upon the system.

CHRONIC IRRITATION OF THE MUCOUS MEMBRANE OF THE STOMACH AND BOWELS.

The largest class of invalids that resort to our mineral fountains for relief are those afflicted with *abdominal irritations*, and especially with *irritations of the mucous coat of the stomach and bowels*.

These irritations are occasionally so masked by a superadded nervous mobility as to conceal their true character from the sufferer, and sometimes from his medical adviser. The disease is far more common in late than in former years. The number of cases at the White Sulphur has been, I am sure, more than triplicated within the last few years. It may be induced by any of the numerous causes whose tendency is to derange the digestive, assimilative, and nervous functions; and is often connected with some indigestion, irregular or costive bowels, with restlessness and unhappy forebodings of impending evils. I have much confidence in the waters in such cases when prudently and cautiously used, aided, if necessary, by proper adjunctive means, and pressed to their full *alterative effects*.

LIVER DISEASES.

Chronic disease of the liver, in some form or other, is a very common disease of our country, especially in the warm latitudes and miasmatic districts. Very many affected with this complaint have annually visited the White Sulphur for the last fifty or sixty years. In no class of cases have the effects of the waters been more fully and satisfactorily tested than in *chronic derangements of the liver*.

The *modus operandi* of sulphur water upon the liver is dissimilar to that of mercury, and yet the effects of the two agents are strikingly analogous. The potent and controlling influence of the water over the secretory function of the liver must be regarded as a specific quality of the agent, and as constituting an important therapeutic feature in the value of the article for diseases of this organ. Its influence upon the liver is gradually but surely to unload it when engorged, and to stimulate it to a healthy performance of its functions when torpid.

The control which this water may be made to exercise over the liver in correcting and restoring its energies, is often as astonishing as it is gratifying,— establishing a copious flow of healthy bile, and a consequent activity of the bowels, imparting a vigor to the whole digestive and assimilative functions, and, consequently, energy and strength to the body, and life and elasticity to the spirits.

For many years I have kept a *"Case-book"* at the White Sulphur, and have carefully noted the influences of the water upon such cases as have been submitted to my management. Among the number are several hundred cases of chronic affections of the liver, embracing diseases of *simple excitement, chronic inflammation, engorgement and obstructions of the biliary ducts,* etc. These cases were treated either with the White Sulphur alone, or aided by some appropriate adjunctive remedy; and, in looking at the results, I must be permitted to express a doubt whether a larger *relative* amount of amendments and cures has ever been effected by the usual remedies of the medical shop. This I know is high eulogy of the water in such diseases. It is considerately made, and is not higher than its merits justify.

When *Scirrhosity* of the liver is suspected, the water, if used at all, should be used under the guards of a well-informed medical judgment; for in actual Scirrhosity, if it be pressed beyond its primary effects upon the stomach and bowels, it is very decidedly injurious. I have known several cases in which death was hastened by disregarding this caution.

For a more full account of the influences of the water in Liver diseases, the reader is referred to the author's work on the "Mineral Springs of the United States and Canada."

JAUNDICE.

This is a form of liver disease in which obstructions prevent the free egress of the bile from the gall-bladder along its natural channels, and hence occasion its absorption into the general circulation.

In cases of jaundice, in which the obstructing cause is inspissated bile, or very small *calculi*, or when occasioned by inflammation or spasm of the gall-ducts themselves, the White Sulphur water, as might be expected from its influence over the liver, is used with the happiest results.

Indeed, the individuals affected with incipient or confirmed jaundice, and whose livers are free from Scirrhus, cannot place too much confidence in the use of the White Sulphur water and *baths*, with the occasional use of mild adjunctive means to aid in its speedy action upon the liver and skin. Thus judiciously employed, and for a sufficient length of time, it invariably proves successful, either in curing the case, or in bringing the system into the condition under which a cure speedily results.

CHRONIC DIARRHŒA.

In *Chronic Diarrhœa*, especially where the mucous coat of the bowels is principally implicated, and still more when the case is complicated with derangement of the stomach and liver, the water is often employed with very gratifying effects.

While the water, properly taken, is a most invaluable remedy in *Chronic Mucous Diarrhœa*, in no other disease are prudence and caution more eminently demanded in its administration, and especially for the few first days of using it. When prudently and cautiously prescribed in such cases, it is not only a perfectly safe remedy, but also eminently curative in its effects. Many of the most satisfactory results that I have ever accomplished by the prescription of the White Sulphur water, have been in cases of *Chronic Mucous Diarrhœa*.

SEROUS DIARRHŒA of *chronic character* requires still greater caution in the early use of the water than the mucous form to which I have been referring; and while the waters, when carefully introduced, constitute a valuable remedy in such cases, they will, if too largely taken, aggravate the worst symptoms of the disease.*

---

* See the details of several interesting cases in the " Mineral Waters of the United States and Canada."

COSTIVENESS.

Habitual costiveness is a state of the system in which the water has been extensively employed; sometimes successfully, sometimes not. When the case depends upon depraved or deficient biliary secretions, much reliance may be placed upon the efficiency of this remedy if it be carried to the extent of fully *alterating* the system.

PILES.

The use of mild laxatives in *hemorrhoids* has long been a favorite practice for their relief. The beneficial effect of the water in this disease is probably to some extent due to its laxative power, but still more, I apprehend, to its *alterative effect* upon the liver, through which the hemorrhoidal vessels are favorably impressed.

DISEASES OF THE URINARY ORGANS.

The White Sulphur waters are used with very good effects in *Gravel;* indeed, they almost invariably palliate such cases, and frequently, in their early stages, entirely cure them.

Incipient calculous affections are relieved by the water pretty much in proportion as it corrects the digestive and assimilating functions, improves the blood, and brings the general economy into a natural type, preparing the kidneys to resist foreign encroachments upon their functions, and to elaborate from healthy blood proper and healthy secretions. Where the affection depends upon *acid* predominance in the fluids, the water never fails to palliate, and often cures the case. Whether or not this water should be preferred to other remedies in *calculous* affections, depends upon the *diathesis* that prevails in the system; and hence the urine should always be carefully analyzed, that we may not act in the dark in such cases.

Chronic inflammation of the kidneys, as well as similar affections of the *bladder* and *urethra*, are often successfully treated by a judicious use of the waters. I have treated numerous cases of *Catarrh* of the bladder successfully by a proper use of the water, and other appropriate remedies in connection with it, always regarding the water, however, as the leading remedy in the case.

Diabetes is a form of disease in which the waters have occasionally been used with excellent effect.

Spermatorrhœa, often painfully implicating the nervous system, and producing extreme debility not only of the sexual organs but also of the general system, is often greatly benefited at these springs.

This disease is generally found complicated with a condition of the *skin* and *glandular organs*, and not unfrequently of the mucous surfaces, that eminently requires the aid of *alterative remedies*. In all such complications the waters are found very valuable as a primary means, preceding and preparing the system for the use of more decided tonic remedies.

FEMALE DISEASES.

In *female diseases*, in their various chronic forms of *amenorrhœa*, or suppressed menstruation, *dysmenorrhœa*, or painful menstruation, *chlorosis*, and *leucorrhœa*, the waters of the White Sulphur have been much employed. When the cases have been judiciously discriminated and were free from the combinations and states of the system that contra-indicate the use of the waters, they have been employed with beneficial results.

CHRONIC AFFECTIONS OF THE BRAIN.

It is only since the introduction of the custom of using the water in its *ungaseous form* (thirty-five years ago) that it has been taken successfully, or even tolerated by the system, in chronic inflammation of the brain. I need, therefore, scarcely apprise my readers that it is only in its strictly ungaseous form that it should be used in such cases, and then in a careful and guarded manner. Thus prescribed, I have, in several instances, found it beneficial.

NERVOUS DISEASES.

Neuralgia, in some form or other, has become a very common disease in every part of our country; and the number that visit the White Sulphur suffering with this *protean* and painful malady is very considerable.

Sometimes this disease exists as a primary or independent affection, but far more frequently as a *consequence* of visceral or organic derangements. Where such is found to be the case, the White Sulphur waters are used with the very best results. As an *alterative*, to prepare the neuralgic for receiving the more tonic waters to advantage, it deserves the largest confidence by those afflicted with this annoying malady.

PARALYSIS.

The number of *paralytics* that resort to the White Sulphur is large, and their success in the use of the waters various. Cases resulting from dyspeptic depravities are oftener benefited than those that have resulted from other causes. In almost every case, however, some benefit to the general health takes place, and sometimes an abatement of the paralysis itself.

BREAST COMPLAINTS.

In *tubercular consumption*, whether the tubercles be incipient or fully developed, the White Sulphur water should not be used. Its effects in such cases would be prejudicial.

But there are other forms of *breast complaints* in which the waters have been found valuable, particularly in that form described as

SYMPATHETIC CONSUMPTION.*

This form of breast complaint is the result of morbid sympathies extended from some other parts of the body, and more commonly from a diseased stomach or liver. The great *par vagum* nerve, common to both the stomach and lungs, affords a ready medium of sympathy between these two organs. In protracted cases of dyspepsia, the stomach often throws out morbid influences to the windpipe and surfaces of the lungs, occasioning *cough, expectoration, pain in the breast,* and many other usual symptoms of genuine consumption. So completely, indeed, does this *translated* affection wear the livery of the genuine disease, that it is often mistaken for it.

This form of disease comes often under my notice at the Springs, and I frequently witness the happiest results from the employment of the water in such cases; and the more so, because its beneficial effects resolve a painful doubt that often exists in the mind of the patient as to the true character of the disease.

BRONCHITIS.

This affection is often met with at the Springs, sometimes as a primary affection of the bronchi, and often as a result of other affections, and especially of derangements of the digestive and assimilative organs. In such *translated* cases, we frequently find the *bronchitis* relieved in the same degree that the originally diseased organs are benefited.

CHRONIC DISEASES OF THE SKIN.

The various chronic diseases of the skin are treated *with much success* by a full course of the White Sulphur waters in connection with a liberal course of *warm* or hot sulphur baths.

---

* For fuller information on this subject, see "Mineral Waters of the United States and Canada," by the author.

There is a chronic form of *erysipelas*, occurring at irregular intervals, and most commonly attacking the face or the extremities, that I have treated with very good success by the White Sulphur water.

RHEUMATISM AND GOUT.

The *rheumatic* and the *gouty* are *habitués* of the White Sulphur. The well-established reputation of the waters in such cases attracts no small number of persons laboring under one or the other of these affections.

The primitive reputation of the water, and that which at an early day directed public attention to its potency, was derived from its successful employment in rheumatism. The reputation thus early acquired has not been lost, but, on the contrary, established and confirmed by its successful use for three-quarters of a century.

In most rheumatic cases, the employment of *warm* or hot *sulphur baths* constitutes a very valuable adjunct in their treatment.

With the sulphur water as a drink, and the use of the hot *tub douche* and *sweating* baths of the same water, this place offers the strongest inducements for the resort of persons afflicted with chronic rheumatism that can anywhere be found.

In proportion as the waters impress the digestive and assimilative organs, they benefit *gout*. As *palliative* in this disease, they are always employed with benefit.

CHRONIC POISONING FROM LEAD

Is very advantageously treated by a full course of the water and baths. Used with sufficient persistency, they may well be regarded as the most reliable remedy to which persons thus afflicted can have recourse, and to such I earnestly recommend a trial of them, the more especially, because the ordinary remedies in such cases are admittedly very unreliable.

SCROFULA.

Sulphur waters have long been held in reputation in the treatment of scrofula. Some of the English physicians have thought such waters superior to any other remedy in scrofula. Dr. Salisbury, of Avon, New York, speaks favorably of his experience of their use in such diseases. In the early stages of scrofula the White Sulphur has often been used with decided advantages, but in the confirmed stages of this disease I do not consider them at all equal in curative powers to some other mineral waters in this region.

MERCURIAL DISEASE AND SECONDARY SYMPTOMS OF LUES.

In that enfeebled, susceptible, and very peculiar condition of the system, often found to exist as the result of a long-continued or injudicious use of *mercury*, and in what is commonly known as the *secondary form of Lues*, the White Sulphur water, when carried to its full alterative effects, displays its highest curative powers. After long experience of the use of the waters in the peculiar forms of disease under consideration, I have no hesitation in saying, that if called upon to designate the particular affection or state of the system in which the White Sulphur water is most certainly efficacious, I would not hesitate to name *mercurial diseases* and *Secondary Syphilis;* because the water in such cases exerts a specific agency, and more certainly brings relief to the sufferer than any other known remedy. This is strong praise of the remedy in these diseases, and nothing but long and successful experience of its value in such cases could induce me to award it.

I have no hesitation in saying to those who are so unfortunate as to be subjects of the diseases embraced under this head, that they have in these waters, when *properly* and *fully used*, in connection with warm and hot sulphur bathing, a reasonable hope of a permanent cure that they cannot have from the use of any other remedy known to the profession. Such cases require *a full use of the waters*, and in every case the cure is obviously hastened by the use of other appropriate means while the water is being taken.

EFFECTS OF THE WATER IN INEBRIATION.

During the whole period of my residence at the Springs, I have been interested with the marked power I have seen manifested by the waters in *overcoming the desire for the use of ardent spirits* in those who had been addicted to their imprudent use. I by no means claim that the waters should be regarded as a specific against either the love or the intemperate use of alcoholic drinks, but simply that a proper use of them is a decided preventive of that feeling of *necessity* or *desire* for the use of strong drinks which drives the inebriate to use them, in despite of his own judgment to the contrary. Or, in other words, that their proper use allays or destroys the aptitude or *nervous craving* for ardent spirits, and to such an extent, that even the habitual drinker and confirmed inebriate feels little or no desire for them while he is properly using the waters.

During my long residence here, I have witnessed hundreds of cases fully justifying the above statement. This peculiar influence of the White Sulphur water depends, *first*, upon the action of the *sulphuretted hydrogen gas* that abounds in it, and which is an active nervine stimulant, and as such supplies the want the inebriate feels for his accustomed alcoholic stimulant; and *secondly*, it depends upon the *alterative* influences exerted by the waters upon the entire organism. While by its alterative power the entire animal structure is brought into natural and harmonious action, there is a consequent subsidence of the *cerebral* and *nervous irritation* which always prevails in the habitual drunkard, the abatement of which enables him to exert a moral power greater than he could before, and sufficient to overcome the lessened demand which his old habit, if he retains it in any degree, now makes upon him.

In the initiatory or forming stage of intemperance, the free use of this water may be much relied upon to modify, or entirely prevent, the *temptation* for strong drink; and even in the confirmed stage, its persevering use may inaugurate a state of the system that will essentially aid the sufferer in overcoming the hurtful habit of intemperance. Indeed, if the habitual drinker can be prevailed upon to use the water properly for some ten days, *to the entire exclusion of alcoholic stimulants*, he will have, for the time at least, but little alcoholic temptation to resist.

Of course, I will not be so misunderstood by any as to suppose that I design even to intimate an opinion that this water is a *sure and permanent cure* for either *absolute* or *threatened* inebriation. All that I intend to assert in this connection is, *that a proper and continuous use of the water will very essentially aid the intemperate drinker* to lay aside the inebriating cup and return to soberness.

The *will* of the excessive drinker must necessarily concur, to some extent, with any effort successfully made for his relief. But while this is so, an auxiliary agent, as innocent in its effects as sulphur water, that can so far satisfy the *nervous cravings* of the votary of strong drink as to give him increased power to resist his morbid habit, while at the same time his general health is improved, well deserves, I conceive, the attention of all who need assistance in this direction.

It would be irrational for the inebriate to expect to be cured of his morbid habit by simply visiting the Springs and drinking its water, however freely, *and at the same time* (which has been the habit of some) to drink freely also of alcoholic liquors. Such a course could be of no service whatever. Stimulants of whatever kind, in such a case, *must be abstained from* while the water is establishing its peculiar action upon the system. This effected, which can ordinarily be accomplished in ten or twelve days, the success of further persistence in the use of the water is *hopeful*, and easily thereafter under the control of the individual who is seeking relief.

USE OF THE WATER BY OPIUM-EATERS.

I am occasionally consulted by distant parties who are apprised of the effects of the water in *allaying the desire for ardent spirits*, whether or not it has the same effects in reference to the desire for *opiates*.

Upon this subject I remark, that my observations of the influences of the water in *assisting* the *inebriate* to discontinue the use of alcoholic drinks, when his *will* assents to such discontinuance, very naturally led me to hope that it might afford similar assistance, under a *like consent of the will*, to the *opium-eater*. But a good deal of difficulty lies in the way of making reliable observations upon this subject. Opium-eaters, even more than excessive drinkers, are indisposed to divulge their morbid propensity to their friends or physician, or to seek through the aid of either to be relieved of their hurtful habit; consequently, while personally I have known hundreds of visitants to the Springs who I was satisfied ate opium to excess, and some to very great excess, nevertheless, I have had but few cases of inveterate opium-eating placed fully under my professional government, with the *single view* of being cured of the habit. Some such cases, however, I have had, in which the sufferers freely and fully communicated to me the fact of their injurious habit, expressed earnest desire to be relieved, and continued during the treatment to exercise all the force of will of which they were capable, to render my advice and prescriptions successful. In one of these cases, entirely successful in its treatment, the person had been in the habit for a long time of using not less than *six grains of morphia* daily.

The space allotted to this notice will allow me only now to say, that in the few cases alluded to, I used the waters very fully, but always *in connection with other means* that I deemed essential,—that the success of the combined treatment was very satisfactory,—that, in my opinion, the influences of the water, *by lessening the nervous craving for opiates*, materially *aided* in the results, and that such results would not have taken place if the waters had not been used. In the cases alluded to, a generous confidence on the part of the sufferer, which led to prompt observance of professional advice, contributed much, I conceive, especially in the commencement of the treatment, to favorable results.

The most that can confidently be said in favor of the use of the waters in such cases—and all that ought to be said—is, that when they are *judiciously used, and in connection with proper adjunctive management* and appliances, they essentially *aid* the opium-eater in dispensing entirely with the use of that drug. I will only add that, in my management of such cases, I have not found it best to *exclude the entire use of the drug* when the patient *first commences the use of the water*, as I advise shall be done in the case of the inebriate.

I have not hitherto published anything upon this subject, and simply from the fact that I am satisfied that the treatment of such cases by the waters, to be successful, requires careful professional management, with appropriate adjunctive means,—that the water is only an *efficient aid*, and not a *specific*,—and that the management necessary in connection with it, to give success, depends too much upon the precise circumstances of each case to justify a broad recommendation, without numerous and essential qualifications.

I have heretofore alluded to some diseases and states of the system in which these waters should *not be used*. In addition to what I have already said upon that subject, I now remark that they should not be used in *scirrhous* or *cancerous* affections, whether *internal* or *external*, nor in *hypertrophy* or *morbid enlargements* of the heart. In either of the cases supposed, their effects, especially their full effects, would be prejudicial.

CHALYBEATE SPRING.

About forty rods from the White Sulphur is a *chalybeate spring*, in which the iron exists in the form of a *carbonate of iron*, the mildest, least offensive, and ordinarily the most valuable form in which ferruginous waters are found.

For the last fifteen years this water has been considerably used by the class of visitors whose diseases required an *iron tonic,* and its effects have realized the rational hopes that were indulged in it.

CHAPTER IX.

SALT SULPHUR SPRINGS.

Situation and Early History—Analysis by Professor Rogers—Applicability of the Waters.

THESE springs, three in number, are about twenty-four miles south from the White Sulphur, in the county of Monroe, and near Union, the seat of justice for that county.

The following is the *analysis* of Professor Rogers of the principal spring:—

Temperature variable from 49° to 56°.

Solid matter procured, by evaporation, from 100 cubic inches, weighed, after being dried at 212°, 81.41 grains.

Quantity of each solid ingredient in 100 cubic inches, estimated as perfectly free from water:—

1. Sulphate of lime...36.755 grains.
2. Sulphate of magnesia.. 7.883 "
3. Sulphate of soda.. 9.682 "
4. Carbonate of lime.. 4.445 "
5. Carbonate of magnesia...................................... 1.434 "
6. Chloride of magnesium...................................... 0.116 "
7. Chloride of sodium... 0.683 "
8. Chloride of calcium... 0.025 "
9. Peroxide of iron, from proto-sulphate............. 0.042 "
10. An azotized organic matter, blended with sulphur, about.. 4 "
11. Earthy phosphates... a trace.
12. Iodine ....

Volume of each of the gases contained in a free state in 100 cubic inches:—

Sulphuretted hydrogen	1.10 to 1.50 cubic inches.
Nitrogen	2.05 "
Oxygen	0.27 "
Carbonic acid	5.75 "

The above analysis applies to the Iodine, or New Spring, as well as to the Upper, or Old Spring, as the following extract from a letter from Professor Rogers to the proprietors will show:—

"I inclose you a list of the ingredients in the Salt Sulphur water, which applies to the New as well as to the Old Spring, the former having rather a smaller amount of saline matter in general, though in some ingredients surpassing the other. It has been very minutely analyzed, and is the first of all the waters in which I was enabled to detect traces of iodine, which it contains in larger amount than the Old Spring, and, indeed, than most of the other waters in which I have been so fortunate as to discover this material."

The Salt Sulphur water is remedial in cases for which strong sulphur waters are successfully used; and especially in cases that require active cathartic operation. While its cathartic effects are more active than those of any other water in the geological region in which it exists, it is neither harsh nor violent; gently clearing the alimentary canal without debilitating the patient, while its activity promotes the general secretions, invigorates the appetite, and promotes digestion. The cathartic effects of the water are so mild and certain that the stomach is not oppressed by it, nor the bowels irritated; but while the alimentary canal is being relieved, the functions of the system assume their physiological type, and the suspended causes of disease are gradually removed.

In the extensive range of diseases dependent upon *visceral* obstructions, the Salt Sulphur is eminently useful; and in that particular form of simple *dyspepsia* in which constipation is a leading and troublesome symptom, I have found it to be signally efficacious.

CHAPTER X.

RED SULPHUR SPRINGS.

Situation and Improvements—Analysis—Adaptation to Diseases, etc.—New River White Sulphur Springs.

The Red Sulphur Springs are in the southern portion of the county of Monroe, forty-two miles south from the White Sulphur.

The following is the result, given in one view, from the analysis of this water by Mr. Augustus A. Hayes, of Massachusetts.

50,000 grains (nearly seven pints) of the water contain, dissolved as gases (grain measure),—

Carbonic acid	1245 grains.
Nitrogen	1497 "
Oxygen	260 "
Hydro-sulphuric acid	86 "
	3088 "

And afford of—

Siliceous and earthy matter	0.70
Sulphate of soda	3.55
Sulphate of lime	.47
Carbonate of lime	4.50
Carbonate of magnesia	4.13
Sulphur compound	7.20
Carbonic acid	2.71
	23.26

Mr. Hayes remarks, that the peculiar sulphur compound which forms a part of the saline contents of this water has never been described, if it has ever before been met with. While in the natural state, and out of contact with atmospheric air, it is dissolved in the water, and forms a permanent solution. Air, acids, and other agents separate it from the water, in the form of a jelly, and alkaline carbonates, alkalies, water, and other agents re-dissolve it. It has no acid action on test fluids, but bears that character with bases, and forms compounds analogous to salts.

Mixed with a small quantity of water, and exposed to the temperature of 80° Fahr., it decomposes, and emits a most offensive odor of putrefying animal matter, with hydro-sulphuric acid gas. It is to this property that the hydro-sulphuric acid in the water is due, and to the oxidation of a part of this compound most of the sulphuric acid found in the water may be referred.

Mr. Hayes remarks, that "chemical experiments do not show the medicinal properties of the substances operated on. But when a substance, the result of delicately-balanced affinities, gives in its decomposition an agent of powerful action on the animal system, we may conclude that it is an active ingredient, if found in a water possessed of high curative powers. I am disposed, therefore," he says, "to consider the sulphur compound in this water as the principal medicinal agent contained in it; although its action in combination with the other constituents may be necessary to produce the effects for which this water is so justly celebrated."

Mr. Hayes, from his chemical examinations, comes to the conclusion that the *red color* of the matter which is deposited on the slabs, etc., is that of moss or lichen, which finds its habitat in the viscid covering produced by the deposition of the sulphur compound.

The peculiar and distinguishing reputation of this water, as a medicinal agent, is for diseases of the *thoracic viscera*, and, by some, it has been considered remedial in confirmed tubercular consumption. Dissenting entirely from this high claim for the water as a remedy in *confirmed consumption*, my observations for many years enable me to award to it decided efficacy in many cases of *irritation* of the pulmonary organs. In sympathetic or translated affections of the lungs, whether that state be occasioned from disease of the digestive or chylopoietic viscera, or be dependent upon the retrocession of some habitual discharge, the water deserves to be regarded as a valuable remedy.

While the Red Sulphur has been considered peculiarly adapted to the cure of pulmonary diseases,—and it is true that it has a beneficial influence in many cases of this kind,—its good effects equally extend to all cases of subacute inflammation, whether seated in the stomach, liver, spleen, intestines, kidneys, bladder, and most particularly in the mucous membrane.

It is also used with good effects in chronic bowel complaints, leucorrhœa, gleet, catarrh of the bladder, and in some forms of uterine derangement.

With this as with other sulphur waters, if the system should be too plethoric, or too much excited, the use of the water should be postponed until the excitement shall be reduced to a proper state. Commence by taking one glass of water at bedtime, and one before breakfast; after a few days, take two glasses at bedtime, and two before breakfast, one at eleven o'clock A.M., and one at five P.M.; this quantity will generally operate freely on the bowels; if it should fail to produce this effect, a little common salt, magnesia, or cream of tartar may be added. If it is desired to act on the kidneys, increase the quantity of water to three or four glasses between a light supper and bedtime, and the same quantity between daylight in the morning and breakfast-time, two glasses at noon, and one or two glasses about five o'clock P.M., taking care to exercise freely after drinking. The most proper periods for using the water are, at night before bedtime, and in the morning before breakfast-time.

NEW RIVER WHITE SULPHUR SPRINGS.

This name is given to a recently improved sulphur spring on New River, in the county of Giles, a few miles southwest from the Red Sulphur. This property has been improved within the last twenty years, for the entertainment of visitors.

The waters of this fountain have not been analyzed, but they belong to the great Sulphur class, so abundantly found in that geological region, and are valuable in such cases as are usually successfully treated by mild Sulphur waters.

These Springs may be reached by stage or private conveyance from the Virginia and Tennessee Railroad, at Newbern or Christiansburg, or from the Red, Salt, or Montgomery White Sulphur Springs.

CHAPTER XI.

SWEET SPRINGS.

Situation and Early History—Improvements—Analysis—Effects of the Waters—Adaptation of the Waters as a Beverage, and as a Bath, etc.

THE Sweet Springs are situated in a pleasant valley in the eastern extremity of Monroe County. They are seventeen miles southeast from the White Sulphur, and twenty-two east from the Salt Sulphur.

These springs were discovered in 1764, before any of the other mineral waters in this section of the State were known. In 1774, they had attracted so much attention as to be analyzed by Bishop Madison, then President of William and Mary College.

The valley, in which the spring is situated, is about five miles in length, and from one-half to three-fourths of a mile in width, and is bounded on the south by the lofty Sweet Spring Mountain, and on the north by the Alleghany. The spring and bath are situated in the lower end of a small hollow or valley, that makes out from the base of the Sweet Spring Mountain, from which the ground gradually swells on either side. Contiguous to the spring is a grove of a few old natives of the forest that have fortunately escaped the axe of the spoiler, which, together with a sodding of grass, give the means of a pleasant promenade in good weather.

The earlier improvements of the place were of a rude but comfortable character; they have now, for the most part, given way to buildings of a high order of architectural merit, and attractive in every respect.

The *bathing-house* is a tasteful and elegant structure; and the separate reservoirs, for the use of ladies and gentlemen, are of ample size, and arranged to give every comfort to the bathers.

The temperature (Bell) of the Sweet Spring is 73° Fahr., the same as that which, in England, by a strange blunder, is called Bristol Hot Well. There is considerable resemblance between the two in other respects, as well in the evolution of carbonic acid as in the earthy and saline matters held in solution. In the Virginia spring, however, iron has been detected; whereas the Bristol Hot Well has none in its composition.

By the analysis of Rowelle, one quart of the Sweet Spring water contains:—

Saline substances in general	12 to 15 grains.
Earthy substances	18 to 24 "
Iron	½ to 1 grain.

The saline substances are sulphate of magnesia, muriate of soda, and muriate of lime, with a little sulphate of lime. The earthy matter consists of sulphate of lime, a small portion of carbonate of magnesia and lime, with a small portion of siliceous earth.

Professor Rogers, in the course of his geological survey of the State, analyzed the waters of the Sweet Spring, with the following results:—

1st. Solid matter procured by evaporation from one hundred cubic inches, 32.67.

A portion of this is combined with water.

2d. Quantity of each solid ingredient, estimated as perfectly free from water, in one hundred cubic inches:—

Sulphate of lime	5.703
Sulphate of magnesia	4.067
Sulphate of soda	2.746
Carbonate of lime	13.013
Chloride of sodium	0.060
Chloride of magnesium	0.136

Chloride of calcium	0.065
Peroxide of iron (sesquioxide)	0.061
Silica	0.075
Earthy phosphate	a trace.

3d. Volume of each of the gases contained in a free state in one hundred cubic inches of the water:—

Carbonic acid	37.17
Nitrogen	1.86
Oxygen	a trace.

Sulphuretted hydrogen, a trace, too small to be measured.

4th. Composition of one hundred cubic inches of the mixed gases rising in bubbles in the spring:—

Nitrogen	71.7
Carbonic acid	28.3

The chief distinguishing feature of this water is the predominance of the carbonic acid (fixed air) which it contains, and it is properly regarded as the best example of the acidulous waters that is found in our country.

The name of these waters is calculated to convey erroneous impressions of their taste, which is like a solution of a small quantity of a calcareous or magnesian carbonate. The excess of carbonic acid gives, however, the water a briskness, productive of a very different effect on the palate from what an imperfect mixture of the earths would produce.

The first effects of the water (due to its temperature and gaseous contents), when drunk, are a feeling of warmth at the stomach, with a sensation of fullness of the head, and some giddiness. Taken at stated intervals in moderate quantity, it will produce a moisture on the skin, and increase the flow of urine. If the stomach be in a good state, it gives additional appetite, and imparts fresh vigor to the system. Its operations on the bowels vary at first; but, after a more protracted use, it will generally be found to increase a costive habit.

The Sweet Spring water is serviceable in the varieties of dyspepsia accompanied by gastrodynia or spasm, with pains occurring at irregular intervals, and heartburn, where the extremities are cold and the skin torpid. In secondary debility of the digestive canal, from the exhausting heats of summer, or in chronic diarrhœa and dysentery, without fever, or not sustained by hepatic inflammation, advantage will be derived by the internal use of these waters.

If much gastric irritation, or evident phlogosis of the liver, be present, with a parched skin and other phenomena of fever, it will be better to premise one or two small bleedings, followed by the use of a blue pill at night, and a tumblerful or two of the water, to which has been added a teaspoonful of Epsom salts, or twice the quantity of calcined magnesia, early in the morning.

The harassing cough to which young persons are occasionally subject, and which often has its origin in an enfeebled state of the stomach, or in scrofulous habits from the enlargement of the bronchial glands, as also the *tussis humoralis* of old people, will all be materially benefited by the use of these waters. The relief afforded in such cases as these has usually given Bristol Hot Well its reputation in the cure of pulmonary consumption.

Females who have become enervated by long confinement, or from nursing their children, and whose constitutions have suffered for want of exercise and fresh air, will be benefited by the use of these waters, internally and as a bath.

In subacute rheumatism, and in neuralgic attacks, the Sweet Spring *bath* is often useful. In the closing stages of acute rheumatism, the patient is sometimes harassed with a lingering irritability of his system, with tenderness, pain, and inability in the diseased joints, attended with slight feverishness, especially toward the close of the day.

In such cases, while hot or warm bathing would be

injurious, the baths of the Sweet or Red Sweet Springs may be resorted to with the best effects. The use of the *spout* in such cases is valuable, by placing the diseased part under the falling water and allowing it to receive the dash for a short time.

A very efficacious way of applying this water to the surface is by *douche,*—the stream being directed to the part in which the disease is situated,—wherever there is " augmented heat and fixed pain, as over the stomach, or liver, or abdomen generally, above the pubes, or on the loins and sacrum; also to the joints, when the violence of inflammation has not yet subsided, nor passed entirely into the chronic state. If the irritation of the stomach forbids the drinking of the water, *douching* the epigastrium would form a good preparative for its use in this way. *Lumbago*, with some evening fever, *chlorosis* or *fluor albus*, with heat and pain at the loins, would be benefited by douching this part.

"The excess of carbonic acid, and the presence of earthy carbonates in the water, make it useful in calculous and nephritic complaints."

As a tonic, in cases of pure debility, they may be used with advantage, always, however, regarding this as an aphorism, that *they are contra-indicated, and should be withheld, in all cases in which there is positive congestion in any of the vital organs.*

The first sensation on immersion in the Sweet Spring bath is a slight shock, which speedily passes off, leaving the bather with the most agreeable sensations while he disports himself in the sparkling pool.

The bath is unsuited to the *paralytic*, and should be avoided by those in whom apoplectic tendencies are threatening.

In using the bath, " the chief points to be attended to are, that the skin should not be moist or cold with perspiration, nor that there shall be general chill, nor the languor that follows excessive muscular action. The stomach also should be nearly empty, or, at least, not actively engaged in its work of digestion." Many persons are injudicious in remaining too long in the bath. From two to eight minutes will embrace periods adapted to every condition, and only the most robust should remain in the last-mentioned time. In a large majority of cases, indeed, in all cases in which there is much general debility, from *two* to *five* minutes, according to circumstances, will embrace the proper periods for remaining in the bath.

RED SWEET SPRINGS.

CHAPTER XII.

SWEET CHALYBEATE, OR RED SWEET SPRINGS.

Their Analysis—Nature and Medicinal Adaptations of the Waters as a Beverage and a Bath—Artificial Warm Baths, etc.

ONE mile north of the Sweet Springs, on the road leading to the White Sulphur, and just within the southern border of Alleghany County, are the *Red Sweet Springs*.

This property, embracing about 1700 acres of land, affords one of the most productive farms in the State,—a very great convenience to a spring establishment in reference to its supplies.

The improvements subservient to the springs are spacious, well designed, and comfortable, and are sufficient for the accommodation of from three to four hundred persons. Among them are well-designed and spacious *bathing-pools* for gentlemen and ladies, each affording a *douche*, from the use of which the bather may often derive most essential benefit.

There are also here ladies' and gentlemen's *bathing-rooms* fitted up for receiving *hot* or *warm baths* of any desired temperature.

There are two medicinal springs at this establishment, the one a few paces below the hotel, essentially the same, both in quality and temperature, with the old Sweet Springs; indeed, it may be regarded as identically the same water. The other, some forty rods, perhaps, above the hotel, is in many respects like it, but containing a much larger quantity of *iron*, which, being abundantly deposited in the form of a red precipitate, has given it the name of *Red Spring*.

The water of the *Red Spring*, which is the characteristic water of the place, and most relied upon both for drinking and bathing, issues from beneath heavy and irregular stone arches, just at the head of a narrow cove formed by a projecting hill on one side, and on the other by large masses of porous stone, probably deposited there from the Sweet Spring water, which once flowed in this direction.

There are here three fountains, separated by narrow stone partitions, but all running into one common sluice. The upper and boldest of these fountains is about two degrees colder than the two lower ones, and evidently contains less of ferruginous matter. The water issuing from all of them is probably two hundred and fifty gallons in a minute.

The water of the *Red Spring* has been twice analyzed, first by Rowelle, and then by Professor Rogers. According to Rowelle, *one quart* of this water contains—

Carbonate of lime	4 grains.
Carbonate of magnesia	3 "
Carbonate of iron	2 "
Silex	1 grain.
Sulphate of magnesia	1 "
Muriate of soda	½ "
Iron combined	1 "
Carbonic acid.	

The following is the result of an analysis by Professor Rogers:—

1st. Solid matter, procured by evaporation from one hundred cubic inches, weighed, after being dried at 112°, 40.76.

A portion of this is combined water.

2d. Quantity of each solid ingredient estimated as perfectly free from water. In one hundred cubic inches:—

Sulphate of lime	14.233
Sulphate of magnesia	3.107
Sulphate of soda	1.400

Carbonate of lime	1.166
Chloride of sodium	0.037
Chloride of magnesium	0.680
Chloride of calcium	0.010
Sesquioxide of iron	0.320

Organic matter in small quantities.
Iodine, a mere trace.

The iron is no doubt dissolved in the water as a carbonate.

3d. Volume of each of the gases contained in a free state, in one hundred cubic inches of the water:—

Carbonic acid	46.19	cubic inches.
Nitrogen	2.57	"
Oxygen	.20	"

Sulphuretted hydrogen, a trace, too small to be measured.

4th. Composition of one hundred cubic inches of the mixed gases rising in bubbles in the spring:—

Nitrogen	62.5
Carbonic acid	37.5

The temperature of the Red Spring water, as it issues from three different heads, is from 75° to 79°. Frequent examinations of this spring with a thermometer induce me to believe that its temperature is slightly variable, never exceeding, however, one or two degrees of variation.

The analyses of the Red Sweet and Sweet Spring waters, by the same chemist, show that they contain essentially the same ingredients, but in different proportions, both the *salts* and the *gases* being more abundant in the former. The chief difference in the medicinal effect of the two waters is probably owing to the larger quantity of *iron* held in solution by the Red Sweet. While the Sweet Spring contains of iron 0.061 grains in one hundred cubic inches of its water, the Red Sweet in the same amount of water contains 0.320, or about four-fifths in excess. This goes, so far as analysis can be satisfactory, to prove its higher tonic power.

The iron in this water exists in the form of a carbonate, held in solution by carbonic acid gas, constituting the mildest, and, at the same time, the most efficient preparation of ferruginous waters.

While the carbonic acid gas in the Red Sweet is 41.10 grains against 37.17 in the Sweet, the carbonates as a whole largely prevail in the latter. Again, while the sulphate of lime is much the largest in the Red Sweet, the sulphates of magnesia and soda, both aperient in their character, decidedly predominate in the Sweet Spring waters. It may be noted that *iodine*, in small quantity, is found in the Red Sweet, and not in the Sweet; but its quantity is doubtless very small, and I am not aware of any peculiar effects of the water that can, with certainty, be attributed to this agent. It may, possibly, exert some beneficial influence as a tonic in combination with the other ingredients. From a review of the analyses of these two interesting waters, as well as from observation of their effects on disease, it would not be very inaccurate to say that the Red is the Sweet Spring water with a strong iron base. But medical men, who should look closely into the peculiarities of remedial agents, will find upon careful scrutiny of these, that the difference in the amount and combination of their materials must modify, to some extent, their therapeutical agency upon the human system, and that, according to the practical object they wish to effect, they should select one or the other of them.

As a general rule, it is fallacious to adopt the analysis of a mineral water as a guide in its administration. Although an analysis, as correct as can be obtained in the present state of chemical science, is an important assistant in understanding the general nature of remedial waters, and aiding in the formation of general conclusions in relation to them, still, actual observation of the peculiar effects of these agents is greatly more satisfactory, and far more to be relied upon. Mineral waters often produce effects upon the animal economy that are not indicated by their analyses, and, in some cases, they produce results that are directly contra-indicated. But, in reference to these particular waters, there seems to be quite a concurrence between the indications afforded by their analyses and actual observation as to their effects.

With both of these lights before us, we are forced to regard the Red Spring water as being more decidedly tonic in its influences upon the system than the water of the Sweet Spring, and somewhat more exciting, too; hence, all the cautions that have been urged in reference to the contra-indications of the use of the Sweet Spring water, apply even with more force as to the use of this.

From the same lights we also learn that, as a very gentle aperient, and a mild and somewhat *less exciting tonic*, the Sweet Springs have the preference, and especially in such cases as do not admit or require the use of chalybeates. The *diuretic* effect is about the same from the use of either water.

These general principles may, to some extent, I hope, indicate the class of cases that will be most benefited by one or the other of these springs. But it must be confessed that the subject is sometimes an intricate one, requiring a full knowledge of the case, with a careful comparative estimate of the powers of the two waters, to decide with certainty under the use of which the patient will be most benefited. There is, however, a large class of cases that will be essentially, if not equally, benefited by the use of either of these waters. To such cases as require the use of the *iron tonics*, the Red Sweet water is peculiarly well adapted, and may be prescribed with great confidence.

Both internally, and as a bath, the Red Sweet waters are adapted to numerous diseases. As a *tonic* in cases of nervous debility, or of general prostration, the result of prior violent disease, they may be used with great confidence. In *dyspepsia, particularly when connected with gastrodynia*, and irregular pains in the stomach, with want of tone in the alimentary canal, they may be advantageously employed. In *Gastralgia*, or nervous dyspepsia, after the force of the disease has been softened down by the use of medicines, or alterative mineral waters, they deserve the highest commendation.

Cases of chronic diarrhœa have been cured by the Red Sweet waters, after other springs, more commonly recommended for that disease, have failed to give relief.

Simple debility of the uterine and urinary functions is very generally benefited by these waters. *Spermatorrhœa*, and that peculiar nervous prostration connected with excessive or improper indulgences, are very happily treated by them, where regard is had to the state of the system in connection with their use. They are profitably prescribed in debility resulting from exhausting discharges, provided such discharges have left no seat of irritation to which general excitement may cause a rapid afflux of fluids with increased sensibility.

Ladies who are laboring under debility from long confinement or nursing,—those whose health has become impaired from want of exercise in the open air, as well as those who have been enervated by *leucorrhœa*, or other exhausting causes, will be greatly benefited by using the water and bath.

In *Neuralgic* affections, unattended with organic lesion or obstruction, this water is used with very general success, and rarely fails to ameliorate or cure such cases.

In speaking of the waters of the *Red Sweet* and *Sweet Springs*, I wish to be understood as alluding to the *baths*, as well as to the internal use of the waters. In a large majority of cases, the bath is, doubtless, the most prominent agent in effecting a cure. Merely *as a bath*, there is probably little difference in the effects of the two springs. The temperature of the Red Sweet is two or three degrees warmer than the Sweet. This, in some cases, might be a difference of importance, and not to be overlooked by the physician or the invalid.

The effects experienced after coming out of these baths, provided the patient has not indulged himself in them too long, are as remarkable as they are agreeable. They differ widely from the effects of an ordinary cold bath. There is an elasticity and buoyancy of body and spirit that makes one feel like leaping walls or clearing ditches at a single bound. This cannot be from the absorption of any of the materials of the water by the cutaneous vessels. The few minutes that we remain in the water, especially the very short time after the stricture of the skin from the first plunge has passed off, forbid such an idea. May it not be owing to a stimulant impression imparted by the carbonic acid gas to the nerves of the skin, and by sympathy extended rapidly over the whole body?

About a mile from the Sweet Chalybeate, and on the same estate, a bold spring, decidedly *sulphurous in character*, issues from under a heavy ledge of rocks. If the surface waters that probably find a way into this spring were carefully excluded, it might constitute a sulphur fountain worthy of notice.

CHAPTER XIII.

HOT SPRINGS.

Effects of the Waters Internally and Externally used — Analysis — Diseases to which they are applicable — Speculations on Thermalization, etc.

The Hot Springs are in the county of Bath, thirty-five miles northeast from the White Sulphur, and twenty-one west from Millborough Depot. Comfortable bathing-houses have been erected for the accommodation both of male and female patients. In each of these houses suitable arrangements are made for taking the *sweat* or *plunge* bath, as may be desired; or for receiving the *douche* when it may be required.

The several baths are supplied with water from separate springs; they range in temperature from 100° to 106° of heat. The effects of these waters in disease prove that they are medicated, though they are considered by many as simple hot water. They are known to contain sulphate and carbonate of lime, sulphate of soda and magnesia, a minute portion of muriate of iron, carbonic acid gas, nitrogen gas, and a trace of sulphuretted hydrogen gas; and, when used internally, some of the consequences are such as we might expect from our knowledge of their constituent parts.

These waters, taken internally, are antacid, mildly aperient, and freely diuretic and diaphoretic. But, when used as a general bath, their effects are very decided. They equalize an unbalanced circulation, and thereby restore the system to its natural sensibility, upon the existence of which their capacity to perform their several functions, and the beneficial action of all remedies, depend. They relax contracted tendons; excite the action of absorbent vessels; promote glandular secretion; exert a marked influence over the biliary and urinary systems, and often relieve, in a short time, the pain caused by palpable and long-standing disease in some vital organ.

They have been analyzed by Professor William B. Rogers. The saline ingredients in one hundred cubic inches of water are—

Carbonate of lime	7.013
Carbonate of magnesia	1.324
Sulphate of lime	1.302
Sulphate of magnesia	1.530
Sulphate of soda	1.363
Chloride of sodium and magnesium, with a trace of chloride of calcium	0.105
Proto-carbonate of iron	0.096
Silica	0.045
	12.778

The free gas consists of nitrogen, oxygen, and carbonic acid gas. It also contains a mere trace of sulphuretted hydrogen.

The heat of the human body, as ascertained by inserting the bulb of a thermometer under the tongue, is about 96°,—sometimes as high as 98°; and these degrees seem to be the same, with little variation, in all parts of the world, neither affected, in the healthy body, by the heat of the torrid nor the cold of the frigid zones. But this, however, relates only to the internal temperature of the body; the heat of the skin is very variable, and, generally, considerably below the degree of animal heat. This arises from the great cooling process of evaporation, constantly going on over the whole surface; its sensibility to all external impressions, and its exposure to the atmosphere, which seldom rises so high as 98°, even in the highest heats of summer.

From a view of these causes, we will easily be led to

perceive why a bath heated to 98° gives a strong and decided sense of warmth to the skin; and a sensation of slight warmth, rather than of chilliness, is felt, even several degrees below this point.

Whenever a bath is raised above the degree of animal heat, it then becomes a *direct stimulus* to the whole system, rapidly accelerates the pulse, increases the force of the circulation, renders the skin red and susceptible, and the vessels full and turgid.

The temperature of the Hot Spring baths, ranging from 100° to 106°, must be decidedly *stimulant*, and the more or less so according to the particular bath employed. It is probable that to their stimulant power we are mainly indebted for their curative virtue. The soothing and tranquillizing effects, which often follow their use, are the result of their sanative influence in bringing the organism into a normal condition.

Hot baths are potent and positive agents. When applied to the human body they are never negative in their influences, but will do either good or harm, according to the judgment and skill with which they are employed.

Their stimulant influences forbid their use in all acute diseases, and they are contra-indicated in such chronic cases as are attended with high vascular excitement, or exalted nervous susceptibility. There are, nevertheless, a large number of *chronic* diseases in which hot bathing constitutes the most rational and the chief reliance of the invalid. But these potent agents should never be prescribed merely for the *name* of a disease, however carefully its nomenclature has been selected. The precise *existing state of the system*, whatever may be the pathology of the disease, ought always to be carefully looked to before a course of hot bathing is directed.

These baths are found eminently useful in most cases of *chronic rheumatism*, and in the various forms of *gout*. In local *paralysis*, occasioned by the use of any of the mineral poisons, or in metastasis of gout, rheumatism, or other diseases, they may be used with good effect. *Chronic bronchitis*, especially if connected with a gouty diathesis; *deafness*, connected with defective or vitiated secretions of the membrane of the ear; old *sprains*, or other painful injuries of the joints, are often much benefited by their use.

Diseases of the Uterine System, such as amenorrhœa, painful dysmenorrhœa, etc., are often relieved here.

In some of the more obstinate forms of *biliary* derangements they are used with happy effects, particularly the *hot douche*, when applied over the region of the liver to relieve the torpor of that organ.

So much has been written on the medical applicability of *thermal waters*, that I have not thought it necessary here to do more than to lay down a few general principles to guide the invalid in their use, and to allude to some particular diseases, for the cure of which these springs are known to be well adapted.

The cause of the high temperature of thermal springs has long been a matter of curious speculation. Some have attributed it to the agency of electricity; but this must be regarded in the light of an ingenious speculation, rather than the result of observation and facts. It is very common now to regard various phenomena as the result of electrical influences, principally, perhaps, because we know the agent to be very potent and pervading, but partly because of our ignorance of the general laws by which electricity is governed. But, whatever the facts may be, there seems to be no proof approximating to a reasonable probability, that electricity is principally concerned in producing the high temperature of thermal waters.

Another theory, and one which elicits the largest amount of credence from scientific men, alleges that "the heat of thermal springs is owing to the central heat of the globe, and that it increases in proportion to the depth from which they proceed." The philosopher Laplace embraced this theory, and it is, I believe, held by most geologists. It is urged,* and, to some extent, is well maintained, that the temperature of the earth increases, as we descend into it, about one degree for every hundred feet; and if the increase continues in this proportion, we should arrive at boiling water at the depth of less than three miles. In proof of this fact, the regular increase of temperature, as *borings* have descended into the earth in the *artesian* well at Paris, now eighteen hundred feet deep, and throwing out, by a subterranean power, an immense volume of warm water, might be cited. But what are we to do with the apparently refuting fact exhibited in the salt wells at Kanawha in West Virginia? Several of these wells have been bored to the depth of *sixteen* or *seventeen hundred feet*, and without any appreciable increase of temperature.

Other theorists suppose that thermal springs owe their temperature to circumscribed volcanoes, and that such springs are a sort of safety-valve to those subterraneous conflagrations. It is well known that an earthquake, or an eruption of a volcano, has often produced a change in the temperature of thermal springs that were even at some distance from the place where these phenomena occurred.

There is still another theory, "that supposes that the heat of these springs is produced by certain processes going on in the interior of the earth, and that these processes are attended with an absorption of oxygen and a consequent extrication of caloric." In the absence of any positive knowledge on the subject, this theory would seem to be sustained by as much proba-

---

* See Professor Daubeny's essay, in the sixth Report of the British Association for the Advancement of Science.

bility as any of the others that have been alluded to. But this is a subject that falls strictly within the province of geology; and the zeal and success with which that science is being prosecuted, afford reasonable grounds to look to its votaries for some elucidation of this curious topic.

CHAPTER XIV.

WARM SPRINGS.

Analysis—Time and Manner of Using—Diseases for which Employed, etc.

THE Warm Springs are in a narrow vale, at the western base of the Warm Spring Mountain, in the county of Bath, fifty miles west of Staunton, and five miles northeast from the Hot Springs. They are among the oldest of our watering-places, having been resorted to on account of their medicinal virtues for more than ninety years. The property was patented by Governor Fauquier to the *Lewis* family, in 1760.

Several of our medicinal fountains claim to have been known and appreciated by the aborigines of the country. In reference to this particular one, there are many tales related by that venerable class, the *oldest inhabitants*, of the discovery and use of its waters by the Indians.

It is a matter of sober history, that very soon after the discovery of the Warm Springs by civilized man, they became celebrated for their curative qualities, in various diseases, as well as for the mere luxury of bathing; and that they were frequented, at much labor and fatigue, by invalids, before any other (save the Sweet Springs) of the valuable watering-places in Virginia were known.

The following is the result of an analysis of a standard gallon of this water by Mr. Hayes, of Boston:—

Sulphate of potash...	1.371 grains.
Sulphate of ammonia..	0.369 "

Sulphate of lime	14.531	grains.
Carbonate of lime	5.220	"
Crenate of iron	2.498	"
Silicate of magnesia and alumina	1.724	"
Carbonic acid	6.919	"
	32.632	"

The virtues of this water are probably owing to its temperature, rather than to any medicinal agents combined with it. The supply of water is very abundant, —estimated at six thousand gallons a minute. For the *gentlemen's bath*, it is received into a room thirty-eight feet in diameter, and may be raised to the depth of six feet. After it has been used, the water is drawn off and the bath fills again in fifteen or twenty minutes. The bathing arrangement for *ladies* is extensive, convenient, and comfortable. This bath is circular, and fifty feet in diameter; surrounded by twenty-two dressing-rooms, with private baths of warm and cold water, and is the largest and most complete establishment of the kind to be found anywhere in our country. Adjoining the gentlemen's bath, a room has been constructed for a cold *plunge* bath, which is plentifully supplied with common spring water, piped from the neighboring hills, of a temperature of from 60° to 65°.

The common practice in the use of the Warm Spring bath is, to bathe *twice* a day, and remain in the water from twelve to twenty minutes each time. In some cases, especially when the bath is used for cutaneous diseases, the patient may profitably remain in for a much longer period, even from half an hour to one hour. As a general rule, and especially for delicate persons, active exercise should be avoided while in the bath, and always, on coming out, the bather should be well rubbed over the whole body with a coarse towel.

The best times for bathing are, in the morning before breakfast, and on an empty stomach an hour before dinner. Where perspiration is required, the bath should be taken in the evening, the patient retiring to bed immediately after.

The diseases for which these baths have been profitably employed are numerous; among them are atonic gout, chronic rheumatism, indolent swellings of the joints or lymphatic glands, paralysis, obstructions of the liver and spleen, old syphilitic and syphiloid diseases, chronic cutaneous diseases, nephritic and calculous disorders, amenorrhœa and dysmenorrœa. Occasionally, *chronic diarrhœa* is relieved. The same may be said of *neuralgia;* but, most generally, we find baths of somewhat lower temperature more beneficial in this disease. In connection with the internal use of the alum waters, these baths will be found very serviceable in the various and distressing forms of *scrofula*. In painful affections of the limbs, following a mercurial course, they are efficacious, and the more so if employed in connection with the internal use of the sulphur waters.

Some precautions should be observed in entering upon the use of these baths, even by those to whose diseases they may be well adapted. The bowels should be open, or in a solvent condition; the state of the tongue should indicate a good condition of the stomach; the patient should be free from febrile excitement, and from the weariness and exhaustion generally the result of traveling in the public conveyances in hot weather. Many commit an error, and occasionally make themselves quite ill, by imprudently plunging into the bath immediately after arriving at the springs, and before they have in any degree become relieved from the fatigue and excitement of the travel necessary to reach them. From such an imprudent course, the bather has little rational grounds to hope for benefit, and is fortunate if he escape without injury.

Timely and properly used, these baths are entirely safe; and for the *luxury of bathing*, are equal, or superior, to any elsewhere to be found.

CHAPTER XV.

HEALING SPRINGS.

Location—Analyses—Therapeutic Action—Diseases for which they may be Prescribed, etc.

THIS medicinal fountain is in Bath County, Virginia, and is one of the *thermal* springs that give name to that county, and for which the chain of valleys, that lie at the western base of the Warm Spring Mountain, is so remarkable. The most southern of the group is the *Falling Spring Valley*, which embosoms the water under notice.

The Healing Springs comprise three separate springs. Two of these are quite near each other, and the third at a distance of perhaps two hundred yards in the same ravine. These springs are beautifully bright and crystalline; and the ever-bursting bubbles of gas, that escape with the water and float in myriads of vesicles upon its surface, impart to it a peculiar sparkling appearance.

Their temperature is uniformly 86° Fahr., nor are they subject to any variation of quantity or quality.

The following is Prof. Aiken's analysis of what is termed the New Spring:—

New Spring, spec. grav. 100030. Temperature 88 deg. Fahrenheit. Water feebly acid to test-paper. One gallon contains—

Carbonate of lime	18.721	grains.
Carbonate of magnesia	1.964	"
Carbonate of iron	.275	"
Sulphate of lime	1.263	"
Sulphate of magnesia	7.392	"

Sulphate of potassa...	2.530 grains.
Sulphate of iron..	.100 "
Sulphate of ammonia..	.234 "
Chloride of potassium...	.253 "
Chloride of sodium..	.288 "
Silicic acid...	1.820 "
Organic acid, probably crenic.............................	.876 "
Carbonic acid..	2.286 "
Sulphuretted hydrogen...	.00010 "
Bromine } a trace of each. Iodine }	
	38.00210 "

The bubbles of gas that rise contain in 100 parts nitrogen gas 97.25, carbonic acid gas 2.75.

The contents of the *Old Spring* are essentially the same, being somewhat less abundant in solid material.

A species of *algæ* springs up luxuriantly in these waters. It is of a dark-green color, and exceedingly delicate and beautiful in structure. The water, when drunk, acts in three principal ways upon the system, to wit: upon the *kidneys*, the *bowels*, and the *skin;* and the relative affinity for each particular organ is correctly indicated by the order of their enumeration. The direction to either is influenced somewhat by the condition of the system and by the manner of using the water. But while it is capable of being directed to either organ specifically, it may be so employed as to exert a general and not less salutary effect over the whole at once. Its simultaneous action upon three great emunctories of the body, with its capacity to be directed specifically to either, constitutes this water a safe and gentle, but at the same time a certain and efficient, depurating agent of the human body.

Bathing, both general and topical, is a valuable and important mode of employing the water, and should not be neglected when demanded by the circumstances of a given case.

The water of the Healing Springs, so far as it is capable of classification, may be regarded, in its general action upon the system, as *alterative and tonic*, both directly and indirectly; but inasmuch as it is an agent *sui generis* in its character, we doubt the correctness of limiting its action by restrictive definitions.

The first employment of these springs, and their earliest manifestation of curative powers, was in *ill-conditioned ulcers* and *intractable affections of the skin;* and hence the significant name they bear.

Scrofula is believed to be amenable to this agent. Recently, several cures of this malady are reported to have occurred under its use.

In *chronic ophthalmic affections*, gratifying results may be anticipated from the judicious use of these springs.

In all the varieties of ulcers and local inflammations treated by this water, a new agent may be employed; it is the topical application of the moss that grows luxuriantly in the baths and streams that flow from them. This has a peculiar effect. When applied to a diseased surface, it becomes painful, sometimes exceedingly so; and yet, upon inspection of the part, its redness has been dispelled, and a new and more healthy action established. When the application has been long continued, the surface becomes blanched and corrugated.

In *subacute rheumatism* these waters have acquired considerable reputation. For the relief of the suffering, and to correct the morbid condition upon which it depends, they may often be employed, both externally and internally, with benefit.

The temperature of the water is not so high as to stimulate this form into the *acute*, nor so low as to endanger the patient by sudden *metastasis*, while both effects are guarded against by its diuretic action, and its tendency to the bowels and skin. In the present instance, as in other cases, where it is desirable to give the water a decided direction to the bowels or skin, appropriate adjuvants should be employed.

In *neuralgia*, a congener of the disease just considered, the water is frequently found to be remedial, and, from its alleviation of the thrilling, piercing pain attendant upon this affection, one of the springs received long since the homely but expressive title of "Toothache Spring." It is to those cases, dependent upon general derangement of the system, resulting from a residence in unhealthy districts of country, or those that have their origin in nervous irritability, or spring from a gouty or rheumatic diathesis, that the water is adapted.

Dyspepsia, that inveterate scourge of the sedentary and thoughtful, not unfrequently finds an antidote in these waters.

For *chronic thrush* or *aphthæ*, the Healing Springs have been employed with success.

I have occasionally sent patients, suffering under *chronic affections of the lining coat of the bowels*, to this water with good effect.

Leucorrhœa, and other kindred disorders of the female, when independent of malignant action, or actual displacement of organs, will often yield to the free internal and external use of the waters.

Some of the diseases of the urinary organs are favorably controlled by these waters; among which may be enumerated *chronic irritation*, with mucous discharges from the bladder. I have had occasion to be pleased with their effects in several such cases.

CHAPTER XVI.

ROCKBRIDGE ALUM SPRINGS.

Analysis—Remarks on Analysis—The Name Alum a Misnomer, etc.—Therapeutic Effects of the Waters—Diseases in which they are employed—Their Excellent Effects in Scrofula, etc.

These springs are situated in the northern part of the county of Rockbridge, on the main turnpike road leading from the town of Lexington to the Warm Springs, seventeen miles from the former and about twenty-one from the latter.

Small reservoirs cut in the rock receive the alum water as it percolates through a heavy cliff of slate-stone. There are five of these reservoirs or springs, all differing slightly from each other, and also differing from themselves at different times, being stronger, and the water also more abundant, in rainy weather.

These waters were analyzed by Prof. Hayes, of Boston, in 1852.

From a gallon of the water he produced the following results:—

Sulphate of potash	1.765
Sulphate of lime	3.263
Sulphate of magnesia	1.763
Protoxide of iron	4.863
Alumina	17.905
Crenate of ammonia	0.700
Chloride of sodium	1.008
Silicic acid	2.840
Free sulphuric acid	15.224
Carbonic acid	7.356
	56.687
Pure water	58315.313
	58372.000

Alum waters are of somewhat recent introduction as remedial agents, and close practical observation is yet a desideratum as to their peculiar therapeutical agency and most appropriate medicinal applicability. These waters certainly possess unequivocal curative powers, and, although their reputation is now high, they are destined to advance still further in public confidence. Experience has fully shown that they are very efficaciously used in many diseases of the skin and the glandular system, and that in *scrofulous* affections they offer new hopes to the afflicted.

But the name *Alum*, applied to these springs, while it is intended to conform to the general spring nomenclature of calling springs after some one of their leading ingredients, is, medically considered, a misnomer, and conveys the erroneous idea that their virtues are owing to the alum they hold in solution.

Chemically considered, they are an *aluminous sulphated chalybeate,* containing, as will be seen from their analysis, many of the best materials that are found in the most valued mineral waters of Europe or this country. The protoxides of iron, sodium, potash, lime, magnesia, and ammonia, together with sulphuric, carbonic, crenic, chloric, and silicic acids, exist in the water in common with alum. Some of these ingredients are found in the most distinguished of the English and German waters, particularly in those of Tunbridge, Harrogate, Leamington, and Aix-la-Chapelle, as well as in the waters of the famous Spa, in Belgium; in those of Passy, and in the celebrated springs of Bagnères, in Garonne; all of which have acquired a world-wide celebrity for the cure of many diseases for which the Rockbridge Alum has been successfully prescribed.

The fact should always be borne in mind by those who are investigating mineral waters, that it is rather to the *compound*, than to any single ingredient of a mineral water, that we are to look for its medicinal efficiency and the scope of its applicability. That alum is an important ingredient in the compound of this water I do not mean to question, but that it is so transcendently important as to give name to the spring is very questionable. It is said that a rose by any other name will smell as sweet, and so will this *aluminous sulphated chalybeate* be just as efficacious under the appellation of Alum. But the real objection to the misnomer lies behind this, and exists in the fact that it is calculated to mislead the uninitiated, in the absence of analysis or careful inquiry. Indeed, I have reason to know that persons have not unfrequently been disinclined to visit the *Alum*, influenced by the name alone, and under the impression that the water, as its name imports, would act as an astringent, and therefore be hurtful to them.

But whether the name be, or be not, the best that could have been adopted, it is now a fixture, established by many years' usage, and is not likely to be changed; and my only object in calling attention to the subject is to enter a caution against persons being misled as to the character of the water from the mere name of the spring.

The immediate effects of these waters, under their full and kindly influences upon the system, are those of a febrifuge tonic; resembling the action of some of our best vegetable medicines of that class; but superior to them, from their specific tendency to the bowels and kidneys.

By their diffusible astringent and tonic powers they resolve the congestions of engorged viscera and remove subacute inflammations; thus releasing and giving activity to the fluids, they fill up the superficial capillaries and veins, and give a full, slow pulse, with a warm surface, and soft skin.

They purge mildly, perhaps, two-thirds of the persons that use them freely. Their action upon the *kidneys* is generally prompt, sure, and sometimes active.

Their action upon the *skin* is secondary, and is the result of their sanative action upon the blood-vessels and internal organs, by resolving inflammation and congestions, — and hence is always to be regarded as a favorable indication in the case.

Experience has shown that these waters are efficaciously prescribed in many diseases of the *skin* and glandular system; *lupus* and other malignant ulcerations of the mouth and throat have been cured by them.

In various chronic affections of the digestive organs they are advantageously used.

They are valuable in *mesenteric* affections, particularly in persons, old or young, of scorbutic tendencies.

In *chronic diarrhœa* they display speedy and happy effects.

Being prompt and active as a diuretic, when judiciously used, they are found beneficial in *chronic irritations*, and *debility* of the *kidney*, *bladder*, and *urethra*.

To several of the affections commonly known as *female diseases* they are happily adapted. In *leucorrhœa*, as would readily be inferred from their composition, they are an admirable remedy; often curing that disease, although it has been a complaint of long standing. In *menorrhagia*, unattended with plethora of the blood-vessels, and with the system in a condition to bear tonics, they may be prescribed with confidence. In *amenorrhœa* and *dysmenorrhœa*, where a phlogosed state of the system does not contra-indicate the use of mineral tonics, they may be used to eminent advantage. In the *chlorotic* condition of the female system generally, and especially when the tendency is great to *paucity* or *poverty of blood*, the waters will be used to much advantage.

In *anæmic* conditions generally, and in cases of debility and loss of tone in the nervous system, they may be administered with confidence.

Bronchitis, when in connection with a strumous

diathesis, may be treated by these waters to advantage; in such cases they will be found to be one of our best remedies.

In *gastralgia*, or nervous dyspepsia, they often act kindly and effectively, by changing the action of the mucous membrane and relieving it of its subacute irritation.

They actively promote the appetite and invigorate the digestive powers.

But it is especially in *scrofula* that these waters have won their highest honors and established a reputation among the best mineral waters not only of this country but of the world. Their happy combination of tonic, alterative, diuretic, and aperient qualities renders them an efficient remedy in many of the ills of humanity; but especially in the various forms of *strumous* disease, and even their worst forms, they merit confidence and deserve praise. In this formidable class of affections, whether exhibited in the hardened and enlarged glands, and in ulcerations in children, in ophthalmic inflammations, in mesenteric indurations, or in its more intense and pervading development of adult life, they have been extensively used, and generally with benefit to the sufferers.

But let me not be misunderstood as intending to convey the impression that they will cure every case of this disease, whatever may be its seat, character, or combination; both my judgment and experience fall short of this conclusion; but they both concur in regarding the remedy as among the best, if not the very best, now known for scrofula.

The Rockbridge Alum, as therapeutic water, is not a negative agent: its effects upon the system are positive, direct, and palpable. It is, in a high sense of the term, a *medicinal water*, capable, when properly directed and applied, of doing great good in a wide circle of cases, or, when injudiciously used, of disappointing hopes and producing injury. It does not belong to that anomalous class of agents of which it is often said "they will do no harm if they do no good." Such being the potent character of these waters, the importance that cases which are to be submitted to their use should be carefully discriminated, and that the water should be employed under the direction of judgment and experience, must be apparent to all.

JORDON ROCKBRIDGE ALUM SPRINGS.

This is the name given to a new *alum spring* just opened in the immediate vicinity of the old Rockbridge Alum, and flowing from the same strata of slate formation that supply the water of the latter spring.

The analysis of this water by Professor William Gilham shows that one gallon of it contains—

Of silica..	2.920 grains.
Of sulphate of alumina...........................	5.689 "
Of sulphate of magnesia.........................	4.666 "
Of sulphate of lime.................................	3.808 "
Of sulphate of protoxide of iron.............	8.398 "
Of sulphate of potash.............................	0.658 "
Of free sulphuric acid.............................	8.858 "

Chloride of sodium, in small quantity, not determined.
Organic matter, not determined.

The analysis of this spring, being very similar to that of the old Alum in its immediate neighborhood, shows its therapeutic applicabilities to be essentially the same as those of the latter water. Besides this alum, there is, near the hotel built on the property, a *chalybeate spring*, which has not been analyzed, but promises to be a valuable water of its class.

There is also attached to this property another spring, known as *Iodine Alum Water*, which possesses valuable medicinal powers, and some peculiar to itself. The water of this spring is not only adapted to the treatment of the various diseases for which other alum waters are used, but also, from its highly alterative composition, to be a reliable remedy in cases wherein those waters would be uncertain or inefficient.

BATH ALUM, Bath, Virginia.

CHAPTER XVII.

BATH ALUM SPRINGS.

Analysis—Diseases and States of the System in which they may be Prescribed, etc.

The Bath Alum Springs are situated near the eastern base of the Warm Spring Mountain, on the main stage road leading from Staunton to the Warm Springs, forty-five miles west of the former and five miles east of the latter place.

The valley in which they arise is an extensive cove, irregularly encircled by mountains, with an unproductive sandy soil, and affords indications of salubrity and healthfulness.

It is only within the last twenty-five years that these springs began to attract public attention as a mineral water; and it is not more than twenty years since the grounds near the springs, now so elegantly and tastefully improved, were a wild and primitive forest. The property is owned by Joseph Baxter, Esq., who gives his personal attention to its management.

The improvements here are substantial and convenient, affording comfortable accommodations for a large company.

The Alum waters issue from a slate-stone cliff twelve or fifteen feet high, and are received into small reservoirs, that have been excavated near each other in the rock. These different springs, or reservoirs, differ somewhat from each other. One of them is a very strong chalybeate, with but little alum; another is a milder chalybeate, with more alumina; while the others are alum of different degrees of strength, but all containing an appreciable quantity of iron.

Prof. Hayes, of Boston, the same gentleman to whom we are indebted for the analysis of several of our mineral springs, has analyzed the waters of the Bath Alum, and renders the following results from his chemical investigations.

A standard gallon (58.372 grains) was the measure of water of the spring known as No. 2, used in his analysis, and showed the following results:—

Pure water	58317.206
Free sulphuric acid	7.878
Carbonic acid	3.846
Sulphate of potash	.258
Magnesia	1.282
Lime	2.539
Protoxide of iron	21.776
Alumina	12.293
Crenate of ammonia	1.776
Silicate of soda	3.150
	54.798
Pure water	58317.202
	58372.000

Mr. Hayes states that when much reduced in volume by evaporation, the excess of acid chars the organic acid present, and alters the composition of the salts.

"In considering the composition of these waters, the protoxide of iron is assumed to be united to the sulphuric acid. The change produced by heating is referred to the action of the crenate of ammonia, and is the same as ordinarily where crenates, free from apocrenates, are naturally contained in a water. When mixed with the soluble salts of silver, and exposed to light, the gray color is entirely distinct from that produced by either apocrenates, humates, or any decomposing matter. When the metallic silver and oxide of iron, resulting from the first action, are removed, the mixture by evaporation continues to afford brilliant scales of metallic silver, until reduced to a small volume.

"The gaseous matter in these waters is a mixture of carbonic acid, nitrogen, and a small proportion of oxygen, and the measure is about one volume of the mixed gases to forty volumes of the water. The carbonic acid is given by weight, so that a uniform expression of acid relation is adopted, and no misconception can arise if the reader bears in mind the fact that carbonic acid has more than twice the acid or neutralizing power possessed by the strongest fluid sulphuric acid."

Dr. Strother, an intelligent physician, who long resided in the neighborhood, thought very favorably of these waters in *scrofulous, eruptive, and dyspeptic affections.* He also bears testimony to their good effects *in old hepatic derangements, chronic diarrhœa, chronic thrush, nervous debility*, and in various *uterine diseases*, especially in the worst forms of menorrhagia, and in *fluor albus*, both uterine and vaginal.

In *chlorotic* females, and in a broken-down condition of the nervous system, often in males the result of youthful improprieties, as well as when the system is *anæmic*, but free from obstinate visceral obstructions, this water promises to be very beneficial.

Its high chalybeate and aluminous impregnation manifests decided tonic and astringent powers, and indicates its adaptation to a number of diseases, such as hemorrhages of the passive character, the profluvia, obstinate cutaneous and ulcerative diseases, and *anæmic* conditions of the system generally, that are unattended with visceral obstructions.

CHAPTER XVIII.

Rockbridge Baths—Cold Sulphur Springs—Variety Springs—Stribling's Springs.

ROCKBRIDGE BATHS.

THIS new Virginia *spa* is situated in the county of Rockbridge, on the stage road from Lexington to the Goshen Depot on the Central Railroad, and about equidistant from the two places.

The waters of these baths are impregnated with iron, and abound richly in carbonic acid gas. There are here two bold springs, furnishing sufficient water for two bathing establishments. The property is handsomely and conveniently improved, and capable of accommodating from one hundred and fifty to two hundred visitors.

As a *tonic* bath, adapted to nervous diseases, general debility, and to that comprehensive class of cases found to be so essentially benefited by tonic bathing,—and especially after the use of alterative mineral waters,—these baths will be found highly efficacious, and are destined to be a favorite resort to a large class of invalids.

They are conveniently reached, either from Lexington or Goshen Depot, by stages running over well-graded roads.

COLD SULPHUR SPRING.

This is a very pleasant sulphur spring, about seven miles east of the Rockbridge Alum, and two miles west from Goshen Depot on the Central Railroad, in the county of Rockbridge.

The water of this spring has not been analyzed. It is distinctly sulphurous in character, however, and has acquired a considerable amount of favor as a medicinal agent. The spring is regarded as a place of useful and pleasant resort by those who visit it.

The general medicinal adaptations of the water are the same as those of the other sulphurous waters of the country, which have been fully noticed in treating of the White Sulphur waters, etc.

VARIETY SPRINGS.

This name has been given to a series of fountains in close connection with each other, in the county of Augusta, seventeen miles west from the city of Staunton, and near the "Pond Gap" Station, on the Central Railroad.

The name *Variety*, applied to them, seems to be well chosen, as there are here not only an alum and a chalybeate spring, and one of the peculiar characteristics of the Healing Spring in the county of Bath, but also several others differing from all these, whose precise character has not been well defined.

These waters have been too short a time in use to have established a definite record of their medicinal virtues or adaptations; nor have any of them, I believe, been analyzed; their favorable location, however, and the variety and promising character of their waters, bid fair to bring them prominently into public notice, and ultimately to induce the erection of such improvements as a growing patronage will demand.

STRIBLING'S SPRINGS.

These springs are in the county of Augusta, thirteen miles north of Staunton, from which they may be conveniently reached by stage-coaches.

For many years this place was valued mainly on account of its *sulphur* and *chalybeate* waters, but within the last few years an *alum spring* of much promise has been opened near the sulphur fountain, and the place now presents the three varieties of *Sulphur, Alum,* and *Chalybeate* to the choice of the visitant.

The SULPHUR SPRING has been long known as a safe and valuable water of its kind, efficacious for the various diseases for which such waters are generally employed.

Professor Campbell, of Washington College, has analyzed this spring, and produces the following results from a *standard* gallon of the water:—

Carbonic acid gas	8.250 cubic in.	3.899 grains.
Sulphuretted hydrogen gas	2.470 "	0.912 "
Sulphate of potassa		0.441 "
Sulphate of soda		0.812 "
Chloride of sodium		0.610 "
Carbonate of soda		1.203 "
Carbonate of lime		5.517 "
Carbonate of magnesia		3.864 "
Phosphate of lime		0.002 "
Silicate of soda		0.253 "
Organic matter		1.229 "
		18.772 "

The ALUM SPRING has also been analyzed by Professor Campbell, with the following results from a standard gallon of the water:—

Sulphate of iron	12.125 grains.
Tersulphate of alumina	16.675 "
Sulphate of potassa	1.324 "
Sulphate of lime	6.877 "
Sulphate of magnesia	3.371 "
Chloride of sodium	0.640 "
Crenate of ammonia	0.630 "
Silica	1.550 "
Free sulphuric acid	9.092 "
Carbonic acid gas	3.575 "
	55.859 "

A comparison of this analysis with that of the Rock

bridge Alum shows a striking similarity, not only in the character of the ingredients contained in the two waters, but also in the relative proportion of such ingredients.

While this water holds in solution a larger amount of *sulphate of iron*, *magnesia*, and *lime*, it contains somewhat less of *alumina*, *potassa*, *sodium*, *silica*, and *ammonia*. The Rockbridge Alum, it will be seen, contains a greater weight of *sulphuric* and *carbonic acid gas*.

While both public and professional opinion of the value of this water is very favorable, there seems, nevertheless, not to have been any considerable amount of careful and practical observation of its peculiar therapeutic effects, in a large circle of cases.

But in the absence of such actual observation of its effects, the essential similarity of the water to the Rockbridge waters, whose virtues and adaptations are now pretty well known, leaves no reasonable doubt of the great value of this spring, and indicates with a good deal of clearness its adaptations to the various forms of diseases so happily treated by the waters which it so much resembles in chemical composition.

CHAPTER XIX.

Rawley's Spring — Massanetta Springs — Jordan's White Sulphur Springs.

RAWLEY'S SPRING.

RAWLEY'S SPRING is situated on the southern slope of the North Mountain, in the county of Rockingham, twelve miles northwest from Harrisonburg, and about one hundred and twenty miles northeast from the White Sulphur.

The Rawley water is a *compound chalybeate*, happily adapted, by the association of its medicinal ingredients, to act as an efficient *tonic* and *alterative;* and its successful administration for many years proves that it possesses curative properties beyond those of an ordinary ferruginous tonic.

The following is Professor Mallet's chemical examination of this water: —

Protoxide of iron	1.3214	grs. per Imp'l gallon.
Protoxide of manganese	.0122	" " "
Alumina	.0514	" " "
Magnesia	.3874	" " "
Lime	.3536	" " "
Lithia (detected by spectroscope)	trace.	
Soda	.3065	" " "
Potash	.0721	" " "
Ammonia	trace.	
Sulphuric acid	.5208	" " "
Chlorine	.0315	" " "
Silicic acid	.8163	" " "
Carbonic acid (combined)	1.5624	" " "
Organic matter (including humoid acids)	.3531	" " "

The gases dissolved are as follows:—

Carbonic acid.................... 7.42 cubic inches per Imperial gallon.
Oxygen........................... 2.07   "   "   "   "
Nitrogen.......................... 4.18   "   "   "   "

The protoxide of iron in the water of the two other springs was determined as follows:—

Smaller fountain................. 1.1777 grains per Imperial gallon.
Upper spring..................... 1.5290   "   "   "

This analysis, showing that the water not only contains *protoxide of iron*, with carbonic acid in excess, but also that it contains *silicic acid, alumina, manganese, magnesia, soda, lithia, ammonia, sulphuric acid, chlorine,* and *potash*, evidences that it is not only tonic but also *alterative* in its powers. It may be hopefully looked to as remedial in chronic disease generally which is attended with low and deficient vital action; and especially in chronic anæmia, chlorosis, hysteria, fluor albus, dyspeptic depravities, passive hemorrhages, nervous diseases, and particularly in a large class of female disorders depending upon uterine derangement, with deficiency of vital force, and, indeed, in chronic maladies generally that are connected with paucity or poverty of blood, and consequent weakness of the general system.

Comparing the natural constituents of healthy human blood with the leading ingredients contained in this water, it is not difficult to account for its adaptedness as an alterative and restorer of that fluid, and for its efficiency as a tonic to the relaxed and enervated system generally.

The writer has had considerable professional experience for many years in the direction of this water for his patients, either as a *primary* or *secondary* remedy in their cases, and the results have been generally very favorable to the agent as a restorative and invigorating tonic.

It is not only as a primary and independent remedy that these waters are valuable. In various diseases of the abdominal viscera, and other affections, in which the primary use of *thermal* and *strong alterative sulphur waters* is required, and is essential to the cure of the case *as a first remedy*, the *subsequent use* of these waters, to finish up the case by restoring the wasted energies of the system long debilitated by disease, is often a matter of the greatest consequence to the patient.

The accommodations at this place have recently been much enlarged, and are now sufficient for the entertainment of four or five hundred persons.

The Rawley Springs are reached in one day from Baltimore, by way of the Manassas Gap Railroad to Harrisonburg, or by the Baltimore and Ohio Railroad via Winchester. The Southern and Western traveler may reach them conveniently from Staunton.

MASSANETTA SPRINGS,

formerly known as "*Taylor's*," are in the county of Rockingham, near the Peaks of Massanetta Mountain, and four or five miles east of Harrisonburg.

These springs have been long known as possessing medicinal virtues, and especially for dyspeptic depravities, and for the cure of *agues* of long standing, and other chronic malarial influences.

The waters are believed to be alkaline, and strongly magnesian. Prof. Rogers, upon a qualitative examination, reports them to contain *chlorine, iron, arsenic, potassium, sodium, lime, iodine,* and *magnesia*.

While the medicinal effects of these waters have not as yet been sufficiently tested to make for them a reliable and extensive therapeutic record, they have, nevertheless, so decidedly evinced curative powers as to cause them to be favorably regarded among the mineral waters of the country.

The proprietors of these springs are preparing to

open them more extensively for public use than heretofore.

JORDAN'S WHITE SULPHUR SPRINGS.

These springs are in Frederick County, Virginia, five miles from the town of Winchester, and one and a half from Stephenson's Depot, a point on the Winchester and Harper's Ferry Railroad. They are situated in a small valley, surrounded by hills of no great altitude. The earth in the vicinity of the springs is blended with slate, very porous, and readily absorbs all the water that falls upon it. Hence it is as remarkable for its dryness as is the neighborhood for its exemption from vapors and fogs. The grounds about the springs are well covered with grass, are sufficiently extensive for pleasant promenades, and, withal, are shaded by a variety of ornamental trees.

The spring, although not one of great boldness, affords in abundance a mild, pleasant sulphur water, of the temperature of 57° Fahr., which is said not to be influenced either in quantity or temperature by wet or dry, hot or cold, weather.

The fountain is inclosed by marble slabs, and shaded by an octagonal structure, supported by large pillars.

These waters have never been analyzed, though they will probably be found, judging from the geological position of the fountain, as well as from the sensible properties of the water itself, to contain less *lime* than many of our sulphur waters, and, therefore, more free from the harsh ingredients imparted by the sulphate and carbonate of that mineral; while they hold in solution the other components usually found in our sulphur waters. If this suggestion be correct, it points them out as peculiarly valuable in gravel and the various chronic diseases of the kidneys, bladder, and urethra.

Medicinally, the water acts as a diuretic and slight aperient. As an *alterative*, it is found to be valuable in the various forms of chronic disease in which sulphur waters are commonly beneficial. Among other diseases, *dyspepsia* and the various gastric derangements have derived benefit from its use. The same may be said of *liver disease, hemorrhoids, disease of the skin,* and *rheumatism*, especially when it proceeds from the use of mercury. Several gentleman have borne very decided testimony to the superior efficacy of these waters in *gout*, and their unirritating quality would seem to point them out as a valuable remedy in that disease.

Physicians of eminence, long familiar with the use of this water, speak in high terms of its efficacy in *jaundice*, and in the *functional* derangements of the *abdominal viscera* generally; also in the various *chronic* affections of the skin; in chronic irritation of the kidneys and bladder; in gleet, and especially in female suppressions, unattended with acute symptoms.

CHAPTER XX.

BATH OR BERKELEY SPRINGS.

Early History—Baths and Bathing-Houses—Medical Properties of the Waters—Diseases for which used, etc.—Capon Springs.

The Berkeley Springs are situated in the town of Bath, Morgan County, West Virginia, two miles and a half from Sir John's Depot, a point on the Baltimore and Ohio Railroad one hundred and thirty miles west of Baltimore and forty-nine miles east of Cumberland.

These springs were resorted to by invalids at a very early period, and had great celebrity throughout the Colonies. Hundreds annually flocked thither from all quarters, and traditional accounts of the accommodations and amusements of those primitive times are calculated to excite both the mirth and envy of the present age. Rude log huts, board and canvas tents, and even covered wagons, served as lodging-rooms, while every party brought its own substantial provisions of flour, meat, and bacon, depending for lighter articles of diet on the "Hill folk," or the success of their own foragers. A large hollow scooped in the sand, surrounded by a screen of pine brush, was the only bathing-house; and this was used alternately by ladies and gentlemen. The time set apart for the ladies was announced by a blast on a long tin horn, at which signal all of the opposite sex retired to a prescribed distance, and woe to any unlucky wight who might be found within the magic circle!

The whole scene is said to have resembled a camp-meeting in appearance; but only in appearance. Here day and night passed in a round of eating and drinking, bathing, fiddling, dancing, and reveling. Gaming was carried to a great excess, and horse-racing was a daily amusement.

Such were the primitive accommodations at the first watering-place that was opened in Virginia, and such the recreations and amusements of our forefathers, about the eventful period that ushered us as a nation into the world.

Berkeley has now extensive and convenient improvements, and a summer registry of from twelve to fifteen hundred visitors.

Although these waters possess considerable medical virtues when taken internally, they have been most celebrated as a *bath;* their pleasant thermal temperature, from 72° to 74° Fahr., in connection with other properties, adapting them, as such, to a wide range of diseases. They have never been accurately analyzed, but the presence of purgative and diuretic salts has been ascertained, though the impregnation is not strong and the amount is uncertain.

Internal Use.—This water is tasteless, insipid from its warmth, and so light in its character that very large quantities may be taken on the stomach without producing oppression or uneasiness. Persons generally become fond of it after a time; and when cooled it is a delightful beverage. It is beneficial in several of the chronic and subacute disorders, such as derangements of the stomach, with impaired appetite and feeble digestion unconnected with any considerable degree of organic disease. Its salutary effects in these cases would seem to depend upon the exceedingly light character of the waters and their gentle alkaline properties, neutralizing acidity and invigorating and soothing the viscera.

In the early stages of *calculous* diseases, attended with

irritable bladder, their free use internally and externally is frequently of benefit.

External Use.—Externally used, these waters are beneficial in the whole class of *nervous disorders* that are disconnected with a full plethoric habit, extreme debility, or severe organic derangements.

In cases of relaxed habit and debility, where sufficient power of reaction exists in the system, the tonic and bracing influences of plunges in this water will be very invigorating.

Persons suffering from a residence in a warm, low, and damp climate, and subject to nervous affections, will be benefited by the use of the baths.

To the various chronic affections of the mucous membranes, especially leucorrhœa, gleet, etc., as well as to that peculiar form of bronchitis which depends upon a relaxed condition of the membranes, with general want of tone in the nervous system, the water and baths are highly beneficial. The same may be said as to local paralytic affections, if unconnected with congestion of the brain, or cerebral tendencies.

In mildly *chronic* or *subacute rheumatism*, the bath has long enjoyed a high reputation. Many intelligent persons who have long been familiar with its use, place great reliance on it in this class of cases.

CAPON SPRINGS.

At the western base of the North Mountain, in the county of Hampshire, seventeen miles east of Romney, and twenty-two northwest of Winchester, whence they may be reached by a well-graded but mountainous road, are the *Capon Springs*. They are situated in a narrow vale not far distant from the Capon River, and surrounded by a rugged and romantic mountain scenery, perhaps unsurpassed in *trossach* wildness by any in Virginia. The region is high and healthy, and the sources of amusement (often of consequence to the invalid), and especially those of trout and river fishing, together with the excitement of the mountain chase, are unsurpassed at any of our watering-places.

The improvements at Capon are extensive, affording accommodation for about seven hundred and fifty persons.

The *bathing establishment* here is well designed and handsome, affording twenty bathing-rooms for gentlemen and seventeen for ladies, with comfortable parlors for the use of the bathers. The baths are made of brick, coated with hydraulic cement. Shower and douche baths, and artificial warm baths, are also supplied.

The spring affords about one hundred gallons of water per minute. The temperature of the water as it flows from the fountain is 66° Fahr.; in the reservoir that supplies the baths, about 64°.

The water is essentially tasteless and inodorous. Except in its thermal character, it cannot be compared to that of any of the springs in our great spring region. It more resembles the waters of the Berkeley than any of our other springs. As a bath and a beverage, it will, when properly directed, be found very useful in a wide range of diseases, especially in *idiopathic affections* of the nervous system, *dyspeptic depravities*, chronic derangement of the mucous surfaces, etc. It has acquired reputation, and I believe justly, as a remedy in *gravel* and other derangements of the urinary organs. It is a valuable water, and is destined to increase in favor with the spring-going public.

The Capon waters have been analyzed by Dr. Charles Carter, of Philadelphia, and their principal medicinal ingredients ascertained to be—

    Silicic acid,   Magnesia,
    Soda,   Bromine,
    Carbonic acid gas,   Iodine.

CHAPTER XXI.

Coiner's Black and White Sulphur—Roanoke Red Sulphur—Johnson's Springs—Blue Ridge Springs—Alleghany Springs—Montgomery White Sulphur Springs.

COINER'S WHITE AND BLACK SULPHUR SPRINGS.

THESE springs are situated at the western base of the Blue Ridge Mountain, on the line between the counties of Botetourt and Roanoke, on the borders of one of the most delightful and fertile regions of Virginia. They are immediately on the line of the Virginia and Tennessee Railroad, and within a mile of Bonsack's Depot, fifty miles west from Lynchburg.

These springs, as a public resort, are a product of the recent rapid spring development in Virginia, having been brought into public notice within the last fifteen years.

My personal observation of their effects in health and disease is too limited to enable me to speak positively of their medicinal peculiarities or powers, and, in the absence of an analysis, prudence restricts me from considering their therapeutic character, except in the light of analogy, and from the experience of their use by a few gentlemen who seem to be well qualified to judge of their powers. From such light, I believe these waters will be found a safe and beneficial remedy in a large class of cases usually successfully treated by the mild sulphur waters that exist in the same general geological region.

ROANOKE RED SULPHUR SPRING.

This is one of the new places of valetudinary and pleasure resorts which the recent ardor for spring improvement has brought to the public view.

It is situated in the county of Roanoke, on the road from the town of Salem to the Sweet Springs, ten miles from the former, and about forty from the latter place.

It is called *Red Sulphur* from the color of its deposits, and from its supposed resemblance, as a medicinal agent, to the old Red Sulphur in the county of Monroe.

The waters of this fountain have not been analyzed, nor have they as yet so far made out their medical record of applicabilities and cures as to enable me to speak of them with such particularity as I could desire.

They are mild and pleasant sulphurous waters, and no doubt will be found well adapted to a numerous class of cases successfully treated by such waters.

JOHNSON'S SPRINGS,

now better known as *Hollins's Institute*, are in Roanoke County, eight miles east of Salem. They are mild and pleasant sulphur waters. I find these springs, by a qualitative analysis, to contain twenty-eight grains of solid matter to the Imperial gallon, consisting of the sulphates of soda and magnesia, with the chlorides of calcium and sodium.

This property is extensively and handsomely improved, and, except during the summer months, is occupied as a female seminary.

THE BLUE RIDGE SPRING

is in the county of Botetourt, and immediately on the Virginia and Tennessee Railroad, seventeen miles east of Salem. The water of this spring is strongly *saline* in character, and very much resembles, both in its composition and its medicinal effects, the water of the Alleghany Springs in the county of Montgomery.

ALLEGHANY SPRINGS.

Alleghany Springs are on the south fork of Roanoke River, in the county of Montgomery, three miles south of the Atlantic, Mississippi and Ohio Railroad, at Shawsville.

In the *nomenclature* of mineral waters, they properly belong to the *class* known as SALINE. In local situation, they occupy a central position, geographically and geologically, of the great mineral range extending from Harper's Ferry in the north, to the Chilhowee Mountains in the south. All along this entire range this *class of waters* is found; varying somewhat in their ingredients, but all essentially belonging to the same general class. Nor is this valuable class of waters found in any other portion of our continent in the same abundance and purity.

The springs that represent the extremes of this extensive geological range are the *Montvale*, in Blount County, Tennessee, and the *Shannondale*, in Jefferson County, West Virginia, more than four hundred and fifty miles apart. In the intermediate space between these extremes, evidences are afforded in various places along the mountains of the existence of similar waters; but their existence in purity and in sufficient quantity for general use has only been demonstrated and brought before the public in the springs of "*Shannondale,*" "*Blue Ridge,*" "*Yellow,*" and "*Alleghany,*" in Virginia, and "*Tate's*" and "*Montvale,*" in Tennessee.

In the class of *saline waters* are comprised those springs that contain a sufficient amount of neutral salts to occasion the marked effects of such agents, and especially purgative operations.

Such waters exert but an inconsiderable effect upon the sanguiferous and nervous systems, their efficacy mainly depending on their laxative and purgative operations, by which the alimentary canal is excited to copious secretions, and the secretory functions of the *liver* and *pancreas* are stimulated to pour out their appropriate fluids; besides, like other mineral waters, they are absorbed, and conveyed through the whole course of the circulation, and are applied in their medical efficacy to the capillary tissues and glandular organs. The sympathy between the digestive canal, upon which they operate primarily, and all the other organs of the body, prepares us for witnessing the happy effects which they often exert upon the latter organs by their direct effects upon the former.

Where no considerable irritation or inflammation exists in the mucous membrane of the stomach and bowels, *saline mineral waters* will be found valuable in relieving congestion or irritation of distant organs: *first*, by copious evacuation of fluids; and *second*, by derivation of blood from them to the superficies of the portal system.* Affections of the head, chest, skin, and joints will often be greatly benefited by their prudent use.

From the absorption of saline matters contained in such waters, and possibly from the force of sympathy from other organs, the secretions of the *kidneys* and *skin* are commonly much increased. Such results, often highly beneficial, generally ensue from doses falling short of the quantity usually taken to produce active purging.

The waters of the Alleghany Springs, like all waters of the saline class, purge mildly or actively, in proportion to the quantity drunk and the peristaltic excitability of the bowels. Simply as a purgative, they are very superior in many chronic diseases to the drugs ordinarily used for this purpose, and principally in

---

* Bell.

this, that the invalid can keep up their action upon the bowels for a number of days without suffering that debility of the constitution and loss of appetite which so constantly occur from a similar course of the ordinary purging drugs.

In small and *aperient* doses, they often act most beneficially on the functions of the *skin* and *kidneys*, and especially if the warm bath and gentle exercise be connected with their use. Administered in the same way, we sometimes witness very pleasant influences from these waters upon the mucous surfaces, as well as upon the serous, synovial, and fibrous membranes. Such results are sometimes witnessed in chronic catarrh, rheumatic affections of the joints, etc.

The Alleghany water has been analyzed by Dr. F. A. Genth, of Philadelphia. He found one gallon, 70,000 grains, to contain—

Sulphate of magnesia	50.884290 grains.
Sulphate of lime	115.294022 "
Sulphate of soda	1.717959 "
Sulphate of potassa	3.699081 "
Carbonate of copper	0.000359 "
Carbonate of lead	0.000569 "
Carbonate of zinc	0.001713 "
Carbonate of iron	0.157049 "
Carbonate of manganese	0.060617 "
Carbonate of lime	3.613209 "
Carbonate of magnesia	0.362362 "
Carbonate of strontia	0.060536 "
Carbonate of baryta	0.022404 "
Carbonate of lithia	0.001679 "
Nitrate of magnesia	3.219562 "
Nitrate of ammonia	0.559412 "
Phosphate of alumina	0.025549 "
Silicate of alumina	0.207399 "
Fluoride of calcium	0.022858 "
Chloride of sodium	0.274676 "
Silicic acid	0.882782 "
Crenic acid	0.001921 "
Apocrenic acid	0.000192 "
Other organic matter	1.999121 "
Carbonate of cobalt	} traces.
Teroxide of antimony	
	183.069321 "

Solid ingredients by direct evaporation gave ...184.072000	grains.
Half-combined carbonic acid.............................. 1.885526	"
Free carbonic acid 5.455726	"
Hydro-sulphuric acid.. 0.001339	"
Total amount of ingredients................... 190.411912	"

The *medicinal effects* of these waters are *mildly laxative* or *actively purging*, in proportion to the quantity drunk and the excitability of the bowels.

Simply as a purgative, they are vastly superior in *chronic disease* to the ordinary drugs of the apothecary: principally in this, that the invalid may keep up their action upon the bowels for a number of days without suffering that general debility or loss of appetite which so constantly occurs from a similar course of the ordinary purging medicines.

In small or *aperient* doses they act kindly and beneficially upon the *kidneys* and *skin*, and especially when gentle exercise is connected with their use.

Administered in the same guarded way, they exert a happy influence upon the *mucous surfaces*, as well as upon the *serous, synovial*, and *fibrous membranes*. Such influences are witnessed in chronic catarrh, mucous diarrhœa, rheumatic affections of the joints, etc.

They both primarily and secondarily exert favorable influences upon the glandular secretions. This is sometimes marked in the relief they afford in *jaundice* and other diseases of the glandular structures.

In *dyspepsia* they have acquired a more established reputation, perhaps, than in any other form of disease, mainly, we presume, from the fact that they have been more extensively employed in this than in any other single form of disease.

Dyspepsia is multiform, both in its causes and its pathology, and hence no one remedy is equally well adapted to all its forms and phases. But as a general remedy, adapted to meet the general want in the various *dyspeptic depravities,* this water occupies a decidedly high position among the most valued remedies in such cases.

I by no means intend to assert that this or any other mineral water, or any article of the apothecary, is an infallible remedy in all *dyspeptic* cases; such a position would be alike extravagant and uncandid. But I fully indorse the truthful results of experience, that such waters are among our best remedies in all such cases; always safe when prudently used, and often effective where the usual remedies of the profession have failed.

If called upon to say in what particular form of dyspepsia these waters may be most relied upon, I would say in cases attended with *mucous secretions*, and which often develop alarming palpitations and other unpleasant neuralgic affections. But I by no means regard their efficacy in dyspepsia as limited to such cases.

In *chronic mucous diarrhœa*, alike common and fatal in our southern latitudes, the prudent use of this water is eminently proper. In all cases of this kind the water should be used in small and frequently repeated doses, and its influence upon the secreting surfaces encouraged by the occasional use of a warm bath when such an adjunct can be commanded. A departure from this rule of prudence as to the quantity of the water to be used, would cause it rather to aggravate than benefit the case.

In disorders of the *kidneys*, threatening calculous deposits, these waters may be looked to as a hopeful source of relief. Their efficacy in such cases may be attributed mainly to the alterative changes they effect in the blood and upon the secretory and absorbing functions, and to their increasing the flow of urine, thus giving an easier passage to the extraneous matter, which, when long retained, proves painful and injurious.

These springs may be conveniently reached from the East or South by railroad, by way of Lynchburg; or from the Southwest by way of Knoxville.

The improvements at the Alleghany are extensive and comfortable, affording accommodation for five or six hundred visitors.

MONTGOMERY WHITE SULPHUR.

The *Montgomery White Sulphur* are springs of somewhat recent discovery and improvement. They are situated on the southern slope of the Alleghany Mountain, in the county of Montgomery, a few miles east of the town of Christiansburg, and at a short distance from the Atlantic, Mississippi and Ohio Railroad, from which to the springs a branch railroad has been constructed by the owners of the springs.

Persons visiting this place leave the Virginia and Tennessee Railroad at the *Spring Depot*, on the slope of the Alleghany, and take the company's railroad, on which, in a few minutes, they are conducted to their destination.

These springs are extensively and conveniently improved, and favorably situated for cool and pleasant summer residence. The waters, being distinctly *sulphurous* in character, and withal a bland and pleasant beverage, will be found adapted to a large number of chronic affections that are known to be advantageously treated by sulphur waters generally. They are somewhat less cathartic, and also less stimulant, than many sulphur waters, and hence may be used with more freedom and with greater safety than such waters, by delicate and excitable persons.

CHAPTER XXII.

Yellow Sulphur Springs—Pulaski Alum Spring—Grayson Sulphur Springs—Holston Springs.

YELLOW SULPHUR SPRINGS.

THESE springs are pleasantly situated in an elevated and picturesque part of the county of Montgomery, and are surrounded by variegated and interesting scenery and a productive and prosperous agricultural country. They are three and a half miles from the Atlantic, Mississippi and Ohio Railroad, at Christiansburg Depot, from which they may be reached on a well-graded road.

The spring rises on the east side of the Alleghany, and not more than sixty feet below the summit level of that mountain, and its waters flow into the North fork of the Roanoke, which is two miles distant. In consequence of the great altitude of the spring, the climate in which it is situated is very salubrious, the air being elastic, pure, and invigorating during the hottest days of summer. The water is clear, transparent, and very cool, its temperature being about 55° Fahrenheit.

The spring is very bold, yielding 3600 gallons a day, sufficient to furnish an abundance of water for *warm and hot baths*, a means of using the water highly advantageous to many invalid visitors. In running over rough channels, as well as on the bottom and sides of the spring inclosure, it deposits a brownish-yellow sediment; a bluish sediment is also occasionally observed, supposed to be a *prussiate* of iron. After standing in an open vessel for some twelve or fifteen hours, it loses its styptic taste, becomes flat, and deposits a small quantity of its characteristic sediment.

The improvements at the Yellow Sulphur Springs are very comfortable; the buildings are pleasantly arranged, and combine elegance with convenience. Many of the rooms, as well as the spring and the pleasure-grounds, are delightfully shaded by magnificent forest trees.

Under the name of "Taylor's Springs," or "Yellow Sulphur Springs," this watering-place has been well known and much visited by invalids, for nearly seventy years. As early as 1810 it attracted considerable attention, and had numerous visitors, especially from Eastern Virginia and North Carolina.

In 1855 it was analyzed by Prof. Gilham, who says he finds one gallon to contain—

Carbonate of lime.	8.642 grains.
Carbonate of magnesia.	1.389 "
Carbonate of protoxide of iron.	0.617 "
Free carbonic acid.	4.680 "
Sulphate of lime.	65.302 "
Sulphate of magnesia.	21.098 "
Sulphate of alumina.	3.176 "
Sulphate of potash.	0.107 "
Sulphate of soda.	0.750 "
Protoxide of iron.	traces.
Phosphate of lime.	0.015 "
Phosphate of magnesia.	0.011 "
Chloride of potassium.	0.097 "
Chloride of sodium.	0.076 "
Organic extractive matter.	3.733 "

While this water is decidedly *tonic*, *diuretic*, and mildly purgative in its action, its peculiar composition gives it also decided *alterative* qualities, to the sanative influences of which, many of its best effects are to be attributed.

From seven to eight tumblers, taken at intervals, will usually create a mild cathartic effect; as a *diuretic* it is active; but its evident range of usefulness will be found in its valuable *tonic* and *alterative* properties. As a beverage it lies lightly and comfortably upon the stomach, when drunk even in large quantities. With many persons, especially on commencing its use, it occasions slight excitation both of the physical and mental system, evidenced by a flushing of the face, a pleasant glow over the body, some increase of the frequency of the pulse, and of the animal spirits.

The *alterative effects* of the water are more certainly obtained by taking it in moderately small, rather than in large, quantities, at each period of drinking;—in quantities that will not provoke very decided operations either upon the bowels or kidneys.

Its tendency to increase the appetite and promote digestion is very uniform.

In *dyspepsia*, the water has sometimes produced highly beneficial effects.

In that class of *female complaints* demanding the use of tonics, it is an efficacious remedy, and has often proved successful.

In diseases of the *skin*, especially in the various forms of *herpes*, it is said to display highly curative powers.

In *old ulcers* it has been found beneficial; obstinate cases of many years' standing have been successfully treated by the water, used both externally and internally, that had for years resisted the efforts of surgery.

In *chronic diarrhœa* it is much relied upon by those who have had most experience in its use. Doctors Edie, Wade, Black, and other intelligent physicians residing in the neighborhood of the springs, and who have often prescribed the water in this class of cases, commend it very highly.

In *general debility*, connected with *nervous prostration*, and unattended with serious visceral obstructions, it will always be found a valuable remedy.

Extensive improvements are now in progress at these springs, and among others a large and commodious *hotel*, which, when completed, will greatly increase their capacity for accommodating company, as well as the comfort of visitors. Such increase of accommodation had become a necessity in consequence of the immense visitation to the place within the last five years.

The altitude of these springs,—upwards of 2000 feet above the sea-level,—the cool and invigorating summer climate with which they are blessed, and especially their well-tested and valuable *tonic* and *alterative* water, adapted, as long use of it has shown, to a wide circle of diseases, must always render them a very pleasant and advantageous summer resort, and fully justify the enterprise of the proprietors in making large additional improvements.

PULASKI ALUM SPRING.

This spring is situated in the northwest portion of the county of Pulaski, ten miles from Dublin Depot, on the Atlantic, Mississippi and Ohio Railroad.

This water has not been analyzed, but it very much resembles, both in its sensible qualities and its medicinal operations, the water of the Rockbridge Alum. It enjoys a high reputation in its neighborhood as a remedy for scrofula, cutaneous diseases, and other affections for which the alum waters of Rockbridge have become celebrated.

GRAYSON SULPHUR SPRINGS.

The *Grayson Sulphur Springs* are on the west side of the Blue Ridge, in the county of Carroll, twenty miles south of Wytheville. They rise on the banks of New River, in the midst of scenery remarkable for its wildness and grandeur,—in a region as salubrious and invigorating as any in our country.

These waters are decidedly sulphurous, and have been found useful in dyspeptic depravities, and the various chronic derangements of the chylopoetic viscera. Their earliest reputation, which has been well maintained, was in the cure of rheumatism. For all chronic diseases of the skin, especially for *salt rheum*, *herpes*, and *tetters*, they will be found efficacious; for chronic forms of liver disease they are well adapted; and I am informed by highly respectable medical authorities that they have displayed the happiest effects in numerous cases of amenorrhœa, and in chlorotic conditions of the female system.

There is, quite near the Sulphur Spring, a good *chalybeate*, which may be used to advantage in many cases; and in nervous affections and female diseases it will be beneficial to drink it moderately, in connection with the sulphur water.

The Grayson waters have been analyzed by Professor Rogers. He shows that in a given quantity of their solid contents there are found—

Soda	4 grains.
Carbonate of magnesia	3 "
Carbonate of lime	8 "
Sulphate of lime	2 "
Sulphate of magnesia	3 "
Chloride of sodium	2 "
Chloride of calcium	3 "
Chloride of magnesium	1¾ "
Sulphate of soda	4½ "

Sulphuretted hydrogen and carbonic acid abound in the water.

HOLSTON SPRINGS.

The *Holston Springs* are in the county of Scott, in the extreme southwestern angle of the State, near the Tennessee line, forty miles from Abington, and thirty east of Rogersville, Tennessee. They are on the bank of the North fork of the Holston River, in a wild and romantic region of country.

One of these springs comes within the thermal range, being 68.5° Fahr., or about fifteen degrees higher than the common springs of the surrounding country. Of the saline contents of the water, the most abundant are *sulphates of lime* and *magnesia*, and the *carbonate of lime, chloride of sodium, muriate of alumina, sulphate of soda; phosphate* and *sulphate of alumina* are found in smaller proportions. It is actively diuretic, and, under favorable circumstances, determining to the skin by mild diaphoresis; with many it is mildly purgative.

The water of the Holston Springs was analyzed in 1842, by Professor Hayden, who reports that he found one wine gallon of the water to contain 41.14 grains of saline matter, consisting of chloride of sodium, sulphate of lime, sulphate of soda, sulphate of magnesia, and carbonate of lime, with traces of alumina.

KIMBERLING SPRINGS.

The *Kimberling Springs* are a series of medicated fountains in the county of Bland, Virginia, twenty-six miles from Wytheville.

Some of this group of springs have been chemically examined by Dr. Tuttle, of the University of Virginia, with the following results:

"The first water examined was strongly impregnated with *alum*, and was found to be free from copperas and other deleterious ingredients. A partial analysis showed this water to contain in an Imperial gallon—

Sulphate of alumina	83.069 grains.
Sulphate of lime	17.514 "
Sulphate of magnesia	14.014 "

"The waters from the Sulphur Springs are as yet diluted by admixture of fresh water from other springs in the immediate vicinity. When access from these shall have been cut off, the strength of the mineral waters will, of course, be increased.

"The mineral contents of an Imperial gallon of the Red Sulphur were found to be as follows:

Sulphuretted hydrogen (some loss having occurred in transportation)	.1737 grains.
Sulphate of lime	2.3169 "
Carbonate of magnesia	1.62 "
Carbonate of lime	.7238 "

Chloride of sodium	.4229	grains.
Carbonate of soda	6.208	"
Carbonate of potassa	.750	"
Silica	.6733	"
Organic matter	2.160	"
A trace of iron.		
	14.8749	"

A gallon of water yielded, on evaporation, a residue of............14.607 grains.

"The Blue Sulphur water was found, on a partial examination, to be very similar in composition and in strength to the Red Sulphur water, an analysis of which is given above."

These waters are favorably spoken of by Dr. A. J. Nye and other persons in the neighborhood of the springs, who have had some experience in their use.

CHAPTER XXIII.

Fauquier White Sulphur Springs—Buffalo Springs—Huguenot Springs—New London Alum Springs.

FAUQUIER WHITE SULPHUR SPRINGS.

The *Fauquier White Sulphur Springs* are in the county of Fauquier, fifty-six miles from Washington, and forty from Fredericksburg.

According to analysis, the water is impregnated with *sulphate of magnesia, phosphate of soda*, and *sulphuretted hydrogen*. Its temperature is 56° Fahrenheit, 10½° Réaumur. It has a strong sulphuric smell, and, the taste being not unlike the odor arising from the yolk of a hard-boiled egg, is not, perhaps, at first very agreeable to the palate of a gourmand. It operates *purgatively* and *diuretically;* the cuticular pores being opened, and perspiration, especially if the weather be warm, flows easily and copiously.

This property was beautifully and extensively improved before the war, and had for many years been a place of large and fashionable resort.

During the war nearly all the buildings were burned, but doubtless arrangements will ere long be made by which this heretofore delightful place will be put in a condition to meet the wants of the spring-going public.

BUFFALO SPRINGS.

The *Buffalo Springs* are situated in the county of Mecklenburg, a few miles south of Dan River, and seven west of the town of Clarksville.

The *analysis* of the water shows it to be a sulphated chalybeate. Its temperature, as it flows from the earth, is 60° Fahr. Its specific gravity is 1.058.

The solid contents obtained by evaporating one wine gallon of the water are found to consist of—

Sulphate of magnesia	8 grains.
Sulphate of lime	3.5 "
Sulphate of protoxide of iron	2.6 "
Chloride of sodium	a trace.
Chloride of magnesium	a trace.
Sulphate of soda	1.3 grains.
Sulphuretted hydrogen gas	0.54080 grains.
Total of solid and gaseous contents	15.94080 "

The first effects produced by drinking the water are a flushed face, a quickened pulse, and some giddiness of the head. These symptoms soon pass off, however, and are followed by an increase of appetite, a healthful glow on the surface, with more or less perspiration. Active diuresis sometimes supervenes, and continues as long as the water is used. Occasionally some slight purging takes place for the first day or two, but, unless the mucous membrane of the bowels was previously inflamed, or very irritable, the protracted use of the water is attended with constipation.

The water is stimulant, and, of course, contra-indicated in all diseases of an acute inflammatory character; as, likewise, in all cases of hemorrhage of the lungs, or acute diseases of the bronchial tubes. As a mere tonic, however, it is not wholly inadmissible in chronic affections of the chest; but it should never be resorted to without satisfactory evidence of the absence of tuberculous disease of that cavity.

The principal morbid states to which it seems to be well adapted are *dropsical* affections, *visceral obstructions*, protracted *intermittent* and *remittent fevers, chronic diseases of the skin, dyspepsia,* convalescence from fevers of every grade and type, *female complaints*, and almost every disease of the pelvic organs in both sexes.

The happy blending of *tonic* and *alterative* powers in these waters constitutes them a valuable remedy in a comprehensive class of cases in which these two important influences are demanded for the restoration of health.

In the sallow or jaundiced condition of the skin common to denizens of warm miasmatic districts, and always connected with hepatic derangements of some sort, these waters will be used with excellent success. In the milder obstructions of the liver, spleen, and kidneys, as well as in obstructions in the lesser glands of the system, and in paucity or poverty of the blood, their employment will be valuable. We should look also for highly beneficial results from the judicious use of the waters in chronic irritation of the mucous coat of the bowels, bladder, or urethra, as well as in that wretched form of disease technically known as *spermatorrhœa*, a legitimate and not unfrequent result of youthful improprieties.

HUGUENOT SPRINGS.

This watering-place is in Powhatan County, seventeen miles above Richmond. It is near the centre of a tract of land granted by the British crown to a body of Protestant refugees driven from France by the repeal of the edict of Nantes, in 1685. Hence its name.

There are two springs here, one mildly *sulphurous*, the other *chalybeate*. The first was analyzed by Professor Rogers, who ascertained that it contained the ingredients usually found in the sulphur waters of the country, but in small proportions. The other spring was analyzed by Professor Maupin, who pronounced it a mild and pure *chalybeate*.

In addition to these medicated springs, there is a *well* from which is obtained a water strongly tinctured both with sulphur and iron. This is used, not only for drinking, but for bathing, its medicinal properties when thus employed being considered valuable.

NEW LONDON ALUM SPRING.

For a number of years it has been known that alum is a constituent part of a rock that is found in large masses near the town of New London, in the county of Campbell, ten miles southwest of Lynchburg. An excavation made several years ago into the ground, penetrating this rock, but with no view of obtaining alum water, the virtues of which were not then appreciated, has, from the percolation of the water through the layers of rock, afforded an alum of sufficient purity to be used by the good housewives of the vicinity for "setting their dyes."

The medical reputation acquired within the last few years by the alum waters of Rockbridge, induced the proprietor of this rock to sink a shaft or well into it, with the hope of obtaining alum water in sufficient quantity to be used for medicinal purposes. His enterprise was crowned with entire success. On penetrating the rock to the depth of sixteen feet, he came to several percolations of water, furnishing a sufficient amount to induce him to suspend further operations and to cut an entrance into the basin, or spring, after the manner of ancient wells, and of sufficient size to admit of easy ingress and egress to and from the fountain.

Three or four glasses a day of this water will operate gently upon the *bowels* of some persons; it decidedly promotes the secretion of the *kidneys;* but its leading operation is that of a generous *tonic* and astringent to the animal fibre, increasing the appetite and strengthening the general system.

It has been analyzed by Professor Gilham with the following results:

"A gallon of water furnished the following mineral constituents:

Sulphate of magnesia .... 12.664 grains to the gal.
Sulphate of protoxide of iron................ 23.456   "
Sulphate of alumina............................ 7.240   "
Sulphate of lime .... 18.672   "
Sulphate of potassa............................ 10.160   "

And, in addition, we have of *free* or uncombined sulphuric acid, 19.976 grains.

Between the analysis of this water and the alum waters of Bath and Rockbridge, whose reputation and adaptations are now pretty well established, there is a similarity, in several respects, so striking as to induce the belief that they are suited to the same general range of disease.

The intelligent physician, acquainted with the peculiar action of the alum waters, and looking to the leading indications afforded by the analysis of this, will not fail to perceive that it is pointed out as a valuable remedy in a large circle of cases that require an *alterative tonic treatment*. It will be found valuable in *scrofula* and in the various forms of *salt rheum*, as such waters invariably are; while its good effects in *chlorosis*, and other female affections unattended with febrile action, may be looked to, we would think, with decided confidence.

In *anæmic* and other conditions of the system demanding the use of *tonic* remedies, this water may be used with excellent effect. In cutaneous and ulcerative affections, in primary nervous diseases, in profluvia, and passive hemorrhages, it will be found well adapted.

CHAPTER XXIV.

SPRINGS OF KENTUCKY.*

Harrodsburg Springs—Rochester Spring—Olympian Springs—Blue Lick Springs—Estill Springs.

HARRODSBURG SPRINGS.

HARRODSBURG SPRINGS are situated near the source of Salt River, and in the immediate suburbs of the town of Harrodsburg. They have been extensively and handsomely improved, and, in the language of Dr. Drake, will in this respect "compare advantageously with any to be found in America or Europe."

Dr. Raymond's analysis shows that *one pint* of the water of the GRENVILLE SPRING contains—

Carbonate of magnesia	2.87	grains.
Bicarbonate of lime	0.86	"
Sulphate of magnesia (crystallized)	16.16	"
Sulphate of lime (crystallized)	11.06	"
Chloride of sodium	a trace.	
	30.95	"

One pint of the SALOON or CHALYBEATE SPRING contains—

Bicarbonate of magnesia	0.43	grains.
Bicarbonate of lime	4.31	"
Bicarbonate of iron	0.50	"
Sulphate of magnesia (crystallized)	27.92	"
Sulphate of lime (crystallized)	10.24	"
Chloride of sodium	1.24	"
	44.60	"

* To Dr. Drake, who was one of the brightest philosophical lights of the profession in America, we are principally indebted for our knowledge of the Springs of Kentucky.

Dr. Raymond could not detect either free carbonic acid or sulphuretted hydrogen gas.

The water of the Grenville Spring is the better antacid,—that of Saloon the better tonic. Indeed, small as the quantity of iron is, it sometimes produces an uncomfortable feeling in the head, which is relieved by drinking at the other fountain. In reference to the *excretions*, the water from both acts upon the bowels, kidneys, and skin. Beyond these sensible effects, it pervades the whole constitution, and many classes of invalids very soon feel a renovation of appetite, strength, and cheerfulness, although its primary effects seem to be sedative, not stimulant.

Dr. Drake remarks that "these waters are very beneficial in chronic inflammations and obstructions of the abdominal viscera, in such cases of dyspepsia as are attended with subacute gastritis, and in almost every kind of hepatic disorder, except when the liver is indurated, and consequently incurable. They are almost equally beneficial in chronic inflammations of many other parts of the system, especially of the serous and fibrous membranes. In tonic dropsies, in rheumatism, and in various affections of the periosteum from febrile metastasis, from syphilis, and from mercury, they have often effected a cure when other means had failed." He also speaks very favorably of their employment in urinary disorders and chronic diseases of the skin. He enjoins caution in their use in pulmonary complaints, and considers them hurtful in vomica, tubercular suppurations, and hepatization of the pulmonary tissue.

ROCHESTER SPRING.

ROCHESTER SPRING is a feeble but constant stream, that bursts out about sixty feet below the summit of a ridge of coarse-grained shell limestone. It so nearly resembles the waters just described, that a detailed account of its waters would be superfluous. It is one mile from Perryville and twelve from Harrodsburg.

OLYMPIAN SPRINGS.

The OLYMPIAN SPRINGS constitute one of the oldest and most noted watering-places in Kentucky. They are situated in Bath County, about fifty miles east of Lexington, on the waters of Licking River, which unites with the Ohio opposite Cincinnati.

There are several springs and wells, which present such differences in their composition that of all the watering-places of the West this has been supposed to afford the greatest variety; but Dr. Drake remarks, "I could not myself detect more than three kinds,— a *Salt and Sulphur*, a *White Sulphur*, and a *Chalybeate*."

The SALT AND SULPHUR WELL contains sulphuretted hydrogen, muriate of soda, carbonate of soda, and perhaps a little muriate of lime.

The WHITE SULPHUR SPRING is situated half a mile from the well. This spring is said to have made its first appearance during the earthquakes of 1811. Its temperature is 59°. Its composition is essentially the same with that of the well just described, but the ingredients of the two vary in their proportions. The quantity of sulphur is greater in the spring than in the well; on the other hand, the spring has but a weak impregnation of muriate of soda compared with the well. The proportion of carbonate of soda is nearly the same in both.

The CHALYBEATE SPRINGS are two in number, and are situated about forty yards apart, and half a mile from the Salt and Sulphur Well. They are simple carbonated iron waters.

The Salt and Sulphur waters, Dr. Drake informs us,

are principally drunk; of these, from one to eight tumblers are taken in the morning. Its diuretic effects are prompt, its action upon the bowels very inconsiderable.

BLUE LICK SPRINGS.

BLUE LICK SPRINGS.—At this place there are several springs, all essentially of one kind,—the *sulphurous saline*. They are situated on the bank of Licking River, twenty-four miles from the Ohio, and on the main road that leads from Maysville to Lexington. From the early settlement of the State until within the last eighteen years, salt was manufactured at this place.

The analysis of the Blue Lick waters by Professor Peter* shows that its gaseous contents consist of sulphuretted hydrogen and carbonic acid; and its solid contents, of the carbonates of lime and magnesia, the chlorides of potassium, sodium, and magnesia, the sulphates of lime and potash, bromide of magnesium, iodide of magnesium, silicic acid, with a small amount of alumina, phosphate of lime, and oxide of iron.

The solid contents of the Blue Lick water are to those of the Virginia White Sulphur, as rather more than nine to two. In the former are sixty-four grains of chloride of sodium, or common salt, to the pint; in the latter, but a small fraction. The first contains about three and a half grains of sulphate of lime, the second about ten grains. The White Sulphur holds in solution, however, sulphates of magnesia and soda, both of which are wanting in the Blue Lick; while in its turn the latter has chloride of potassium, and sulphate of potash and bromide of magnesium, which are not in the former. The quantity of sulphuretted hydrogen in the Blue Lick is double that in the White Sulphur. Iodide of magnesium is found in both.

---

* Mineral and Thermal Springs, by Dr. John Bell.

The medical virtues of the Blue Lick water are those of a *saline sulphur*, and are analogous to, but more active than, the Olympian Spring water. It acts freely as a diuretic, but only occasionally as a purgative. It may be used with advantage in nearly all the chronic diseases in which the sulphur waters already described have been recommended. The water employed as a bath can be very properly connected with its internal use.*

ESTILL SPRINGS.

ESTILL SPRINGS, in Estill County, are sulphurous waters. There are two springs here, called *White* and *Red* Sulphur.

The White Sulphur contains 0.09 per cent. of solid contents, the Red, 0.04 per cent., consisting in both cases of carbonates of soda, lime, and magnesia; sulphates of lime, magnesia, and soda; chlorides of sodium, calcium, and magnesium; with hydrosulphate of soda, and a trace of carbonate of iron.

---

* Between the *Blue Lick* and the famous Harrogate Springs, in the North of England, there is a striking similarity; and from some personal knowledge of both these waters, I have no hesitancy in expressing the opinion that the waters of the Blue Lick are well adapted to the same class of cases for the treatment of which the waters of Harrogate have been long celebrated.

CHAPTER XXV.

MINERAL SPRINGS OF OHIO AND INDIANA.

OHIO WHITE SULPHUR.

OHIO WHITE SULPHUR.—Near the geographical centre of Ohio, in the county of Delaware, and immediately on the west bank of the Scioto, surrounded by a country broken, hilly, and beautifully picturesque, arises the *Ohio White Sulphur*. The Scioto is here a rippling, rapid stream, hastily flowing and fretting over beds of boulder rocks, and skirted, for many miles above and below the spring, by slopes or banks of considerable elevation, which gently spread out into undulating table-lands, charmingly interspersed with valley and hill, and blessed with an atmosphere free from malarious influences at every period of the year, and as salubrious as is found in our high mountain ranges.

Under the name of *Hart's Spring*, this place has been known for its mineral waters for more than thirty years. The circumstance that led to its improvement as a spring property by Mr. Hart, its former proprietor, is worthy of note. He had visited the White Sulphur Springs in Virginia, for the relief of a complicated stomach and liver complaint; returning to Ohio cured of his disease, his attention was called to this artesian sulphur fountain, and upon examination he found its waters so strikingly to resemble those of the Virginia Spring as to induce him to purchase and improve it in view of its medicinal value.

The property was subsequently owned by Mr. A. Wilson, of Cincinnati, who erected many new buildings, and otherwise extensively improved the property. The water of this spring is sulphurous, abounding principally in the *sulphates of lime* and *magnesia*, with *chlorides of the same salts*, and with some *oxide of iron*.

These springs for a few years were extensively visited, but now, from some cause or other, have ceased to be kept as a public watering-place. In natural scenic beauty they are excelled by but few of our watering-places.

MINERAL SPRINGS OF ADAMS COUNTY, OHIO.

These springs are thirty miles from Portsmouth, and thirty-six from Ripley.

A qualitative analysis of these waters shows them to contain 120.35 grains of solid matter to the gallon, consisting of chloride of magnesia, sulphate of lime, carbonate of lime, chlorides of calcium and sodium, and oxide of iron, with traces of iodine.

These springs are of comparatively recent resort, but have been used advantageously in dyspeptic depravities, liver complaints, and chronic irritations of the abdominal viscera generally. They have also been successfully employed in disorders of the kidneys, female obstructions, rheumatism, and chronic diseases of the skin, as well as in dropsical effusions.

These springs may be conveniently reached by coaches from Portsmouth, Ripley, or Manchester, on the Ohio River.

The YELLOW SPRING is in Green County, two miles west of the Miami River, and sixty-four north of Cincinnati. Dr. Drake states that "it is a copious and constant fountain, that issues between strata of arenaceous limestone, and thus has geological characters perfectly identical with the Chalybeate Springs of the Olympian valley in Kentucky." The temperature of the water is the same as that of the other springs of the neighborhood, 52° Fahr. The water is beautifully transparent, with a slight ferruginous taste, and is said to resemble in its composition the other limestone springs of the country, with the addition of the carbonate of iron.

Dr. Drake informs us "that its water is *diuretic* and slightly *laxative*, if it can be considered as having this effect at all with any uniformity." He considers the water rather restorative than curative, and as such it is valuable for convalescents. He regards it as a pleasant tonic, and hence valuable in cases of debility, or exhaustion following previous violent attacks, and in nervous disorders.

The WESTPORT SPRING.—It arises (Dr. Bell) "in the bed of Deer Creek, a tributary of the Scioto River, from a vast bed of clay-slate, which for many miles forms the bed of the creek." It is a bold fountain, yielding some twenty gallons of water a minute. It belongs to the *saline* class, and contains sulphate of magnesia and iron; the latter being held in solution by carbonic acid, which gives the water a lively and sparkling appearance as it rises to the surface. The water is said to be mildly cathartic. It will, doubtless, be found valuable in dyspepsia, gastralgia, and a numerous class of functional derangements of the chylopoetic viscera.

FRENCH LICK SPRINGS, INDIANA.

French Lick is situated in a beautiful valley tributary to that of Lost River, about the centre of Orange County, ten miles from Paoli, and eighteen from Orleans, on the New Albany and Chicago Railroad.

There are more than a dozen of these, but all seem to be derived from three parent springs, which are situated within an area of half an acre, but which exhibit some difference in constitution.

FRENCH LICK SPRING.

Pluto's Well, as it is termed, is remarkable for the production of a large volume of strong sulphur water. Its constitution is as follows, according to the analysis of Dr. Rogers, of Madison, Indiana:—

Free Gases in Wine Gallon.

Sulphuretted hydrogen	25.5 cubic inches.
Carbonic acid	15 " "

Salts in Wine Gallon.

Chloride of sodium	140.54 grains.
Chloride of calcium	5.35 "
Sulphate of lime	60.59 "
Sulphate of magnesia	18.12 "
Sulphate of soda	22.37 "
Carbonate of magnesia	1.59 "
Carbonate of lime	6.95 "
Carbonate of iron and alumina	a trace.
Loss	.54 "
Total of salts	256.00 "

All these waters have about the same general medicinal effect. They are alterative and tonic when moderately used; in larger quantities are hydragogue eliminators, acting upon the bowels, kidneys, and skin, without, however, producing the irritating effect which this class of agents usually induces when prepared by means of a pharmacy less perfect than that of nature. They are especially adapted to the treatment of diseases of the skin, dyspepsia, constipation, chronic inflammation of the various mucous surfaces, scrofula, rheumatism, and, in fine, may be beneficially used in all chronic affections where a tonic and alterative effect is required.

CHAPTER XXVI.

MINERAL SPRINGS OF MICHIGAN AND WISCONSIN.

THE ST. LOUIS MAGNETIC SPRING is an artesian well, located in St. Louis, Gratiot County, State of Michigan. It was undertaken with the idea of boring for salt, and was stopped at the depth of two hundred feet, when a flow of water of two hundred and eighty gallons per minute had been struck. Attention was first called to the peculiarity of this spring by observing the strong electrical condition of the tubing of the well, through which the water flows. It was noticed that this tubing would attract and hold small pieces of iron and steel. Pieces of such metals were then laid in the trench that conveyed the surplus water away, and it was found that they became magnetized in a day or two by the action of the water. These facts drew attention to the water as a therapeutic agent of probable value, and to an extensive use of it in cases to which it was supposed to be adapted.

Dr. Samuel P. Duffield, Professor of Chemistry in Detroit Medical College, has analyzed this water, and obtained the following result, calculated on the Imperial gallon. Specific gravity, 1011.

Sulphate of lime	66.50 grains.
Silicate of lime	6.72 "
Chloride of lime	a trace.
Bicarbonate of soda	106.40 "
Bicarbonate of lime	69.40 "
Bicarbonate of magnesia	17.50 "
Bicarbonate of iron	1.20 "
Silica free	2.88 "
Organic matter and loss	2.00 "
Total constituents	272.60 "

Bicarbonates.. 194.62 grains.
Free carbonic acid in gallon........................... 6.21 "
Sulphuretted hydrogen.................................... traces.
　　Total mineral in one gallon...................... 279.60 "

As regards the *rationale* of the magnetic state of this and other similar waters said to exist in the same geological district, it may be observed that they are all the result of artesian borings of the earth for considerable distances, in search of salt. These *wells* are situated along the margins of the great salt and gypsum belts, whose waters hold largely in solution the salts of lime and sodium. It is well known by those who work in salt wells, and to all *scientists*, that when the earth is penetrated to any considerable depth by boring, and iron tubing is introduced into the depth, saline waters flowing through the tubing will possess in a greater or less degree some *magnetic or allied electrical action*. It is caused by *terrestrial magnetism*, which is imparted to the iron tubing, and from it to the water flowing through the tubes. And thus the water becomes the *conductor* of terrestrial magnetism.

This water is strongly alkaline, and as such will prove valuable to a numerous class of chronic disorders, such as rheumatism, kidney, bladder, and other diseases that are known to be successfully treated by waters of this class. But it may well be doubted whether magnetism or electricity can impart to water *molecular agency and cause the molecular changes to be of a permanent nature*. Such results cannot be demonstrated, and there is nothing in all knowledge that is deductive to show how such permanent effects can be accomplished by such agency. Hence it is difficult, without an entire change in existing pathological and therapeutic views, to perceive the *rationale* of ascribing to such agency the various cures said to have been effected by this water, while the admitted efficacy of its therapeutic salts is overlooked or forgotten.

SPRINGS IN WISCONSIN.

BETHESDA is a strong *alkaline* spring recently brought into notice as a therapeutic agent, in the town of Waukesha, eighteen miles distant from Milwaukee.

The following is the analysis of this spring, made by Prof. C. F. Chandler, of Columbia College, New York:—

In one U. S. or wine gallon, of 231 cubic inches, there are—

Chloride of sodium	1.160 grains.
Sulphate of potassa	0.454 "
Sulphate of sodium	0.542 "
Bicarbonate of lime	17.022 "
Bicarbonate of magnesia	12.388 "
Bicarbonate of iron	0.042 "
Bicarbonate of soda	1.256 "
Phosphate of soda	a trace.
Alumina	0.122 "
Silica	0.741 "
Organic matter	1.983 "
Total	35.710 "

This water has been used with marked and excellent effect in numerous cases of diabetes, and in chronic irritations of the kidneys and bladder. Its judicious use will doubtless be found valuable in curing various kidney depravities, and in correcting uric acid predominance in the blood, that often lead to the formation of calculus. Some medical men who have prescribed it, think it decidedly curative in Bright's disease of the kidneys. That its use would be valuable in the early stages of that formidable disease, before positive degeneration of the kidneys takes place, is very probable. Indeed, its efficacy in the early stages of *albuminuria* has been satisfactorily shown from its use.

CHAPTER XXVII.

MINERAL SPRINGS OF TENNESSEE.

White's Creek Spring—Robertson's—Winchester—Beersheba—Montvale—Tate's—Lee's—Sulphur and Chalybeate—Alum Springs—Warm Spring on the French Broad.

THE same great Appalachian chain of mountains that extends through Virginia and West Virginia, and affords what is there known as the "Spring Region," continues its course southwesterly through the State of Tennessee from the northeastern to the southwestern border of the State, gradually losing its elevation as it goes south, until, finally, in Alabama, it sinks into the alluvial plains that extend to the Gulf of Mexico.

This extensive mountain range, or rather series of mountains, running on the same parallel, is called in Tennessee the Cumberland range, and divides East Tennessee from Middle Tennessee.

On the southern border of the State, for nearly two hundred miles in length, is the great chain of the Blue Ridge mountains, a continuation of the same lofty range that in Virginia separates the *Great Valley* from *Eastern Virginia*.

In Tennessee, this range of mountains is on the line between that State and North Carolina, South Carolina, and Georgia. Both of these great mountain ranges afford essentially the same geological characteristics in Tennessee that they do in Virginia; and on their slopes, and near their base, in the latter as in the former State, mineral springs of various qualities and strength are known to exist. But as yet in Tennessee few of these springs have been improved and made places of resort for the invalid, or the general public; nor have they yet, as a general thing, made out a satisfactory record of their precise quality or medicinal applicabilities.

The *saline* and *sulphurous* and the *carbonated iron* waters are those most frequently met with in this State. I proceed to mention those that have been introduced to public notice as places of valetudinary or pleasure resorts.

WHITE'S CREEK SPRING is twelve miles from Nashville. It is held in high estimation by many, and is considerably resorted to. It contains *sulphur*, *iron*, and *magnesia*, the former in large proportion. In *cutaneous* disorders and *calculous* affections it has been much praised for its curative powers.

ROBERTSON'S SPRINGS belong to the class of *saline* waters. They are twenty miles from Nashville.

WINCHESTER SPRINGS are four miles from the pleasant town of Winchester, in Franklin County, on the Nashville and Chattanooga Railroad, seventy miles from Nashville, and fifty from Chattanooga.

There are here, in close proximity, *four* different springs,—Red and White Sulphur, Chalybeate, and *Freestone*. These springs enjoy considerable celebrity and patronage, and are well worthy of attention as a place of both healthful and pleasurable resort.

In the same neighborhood, and but *four* miles distant, other springs have been discovered, called ALLISONA SPRINGS. They resemble the Winchester Springs in quality, and promise to be of equal medicinal value.

BEERSHEBA SPRINGS are on the summit of one of the spurs of the Cumberland Mountain, in the county of Grundy, about twelve miles northeast from McMinns-

MONTVALE SPRINGS, Tennessee.

ville. They have come into notice as a watering-place within the last fifteen years.

The water is a saline chalybeate, and is regarded as a valuable tonic alterative.

These springs have been tastefully and conveniently improved for the accommodation of from four to five hundred persons.

The scenery surrounding the Beersheba Springs is both beautiful and picturesque, and remarkable alike for its extent of range and its wild and romantic prospects.

There are here some fifteen or twenty elegant cottage residences, belonging to and generally occupied by wealthy families of Nashville and other parts of the Southwest.

The society assembled at the place during the summer is always select, elegant, and cultivated, and this, in connection with the value of the waters and the salubrious character of the atmosphere, makes *Beersheba* a desirable summer retreat.

Through the entire circuit of East Tennessee, as bounded by the Cumberland range of mountains on the north and the Blue Ridge on the south, mineral waters are abundant, and some, particularly of the *saline* and *chalybeate* character, have been demonstrated to be of excellent character.

MONTVALE SPRINGS are in Blount County, twenty-four miles south of Knoxville. They belong to the *saline* class.

The analysis of these waters, by Professor Mitchell, shows that they contain in one gallon of water—

Chloride of sodium	1.96
Sulphate of magnesia	12.00
Sulphate of lime	74.21
Sulphate of soda	4.51
Carbonate of lime	13.26
Carbonate of iron	2.40

They also show traces of potash and organic matter, with an excess of carbonic acid.

The *Montvale* are valuable waters, and very favorably represent the class to which they belong. In many of the *dyspeptic* depravities, and generally in the chronic disorders of the abdominal and pelvic viscera, they are used with great success.

They enjoy considerable reputation in the cure of *chronic diarrhœa*, a disease very common and very fatal in our extreme Southern latitudes. In the summer of 1854 the author spent several weeks at Montvale, and witnessed the operation of its waters in quite a number of cases of this disease. In those in which it was used in quantities but slightly provocative of increased operations from the bowels, and in which a guarded forbearance in diet and general living was observed, it proved eminently useful, and especially in cases connected with and kept up by depraved biliary secretions; while, on the other hand, those who used the water in full purgative doses derived no benefit, and some were injured. The best article in the *Materia Medica* may be so misused as to render it inert or injurious, and the invalid at this or at any of the mineral springs should remember that it is not, as many seem to suppose, to *drink and be healed*, but *so to drink* as to secure the proper and sanative effects of the agent.*

The waters of the Montvale more resemble those of the *Alleghany* Springs in Virginia than any other with which I can compare them.

TATE'S SPRINGS are in the county of Granger. They are *saline* waters, and are very like those of *Montvale*, but hold in solution a larger amount of iron.

LEE'S SPRINGS are twenty miles east of Knoxville.

---

* See account of *Montvale Springs*, by J. J. Moorman, M.D., published in 1855.

There are here two *sulphur* springs and a *chalybeate* spring. The sulphurs are good waters of their class; the chalybeate is pure and strong, and superior to many waters of its kind.

At the town of Rutledge, in Granger County, is a very strong sulphurous spring, and near *Bean's* Station, in the same county, are several beautiful fountains of sulphur water, abounding in red and white deposits.

ALUM SPRINGS.—I have examined the waters from an *alum spring* found near Rogersville, in Hawkins County, which compare favorably with any alum waters that are known.

WARM SPRING.—On the French Broad River, near the North Carolina line, there is a warm spring of 95° Fahr. issuing from the bank of the river.

CHAPTER XXVIII.

SPRINGS OF NORTH CAROLINA.

Warm and Hot Springs of Buncombe—Shocco Spring—Jones' White Sulphur and Chalybeate—Kittrell's Springs.

NORTH CAROLINA is not remarkable for mineral springs. The most noted are the—

WARM AND HOT SPRINGS OF BUNCOMBE.—These thermal fountains arise on the western bank of the French Broad River, and so near the stream that in times of high freshets they are overflown by its waters.

The fountains are three in number, and vary in temperature from 94° to 104° Fahr.

Professor Smith obtained the following results from analyzing three quarts of the water:—

Muriate of lime and magnesia	4	grains.
Sulphate of magnesia	6	"
Sulphate of lime	41.05	"
Insoluble residue	2.05	"
Loss	1	"
	27.10	"

Equal to 4.66 grains in a pint.

This water lies lightly upon the stomach, and is often used by visitors to the extent of three quarts, or even more, in the course of the day. In such doses, it is said to excite active purgation when first used, but after a few days it ceases to have any active effect.

As a *bath*, these waters have a wide and appropriate applicability. The bath of 94° will very generally be found safe and salutary for most persons. Those of higher temperature should be used with caution, and with a prudent reference to the nature of the disease and the state of the system at the time of their use. As stated when treating of the Hot Springs in Virginia, hot baths are potent and *positive* agents; they are revolutionary remedies, and to be used safely and successfully must be used with wise discrimination. They are unsuited to persons in ordinary health, and to all acute or subacute cases, but admirably suited to many cases of obstinate chronic diseases, especially to chronic rheumatism, palsy, and other cases depending upon obstinate obstructions and loss of vascular and nervous energy.

An able writer upon baths adopts the following decision as to their temperature, which may well be made a fixed rule to determine the import of language when we speak generally of the temperature of baths:—

1. The cold bath............from............ 33° to 60° Fahr.
2. The cool bath................ " .... 60° to 70° "
3. The temperate bath......... " .... 75° to 85° "
4. The tepid bath............... " .... 85° to 92° "
5. The warm bath............... " .... 92° to 98° "
6. The hot bath.................. " .... 98° to 112° "

He remarks that "the only upward limit of the hot bath is that of tolerance by the living body immersed in it. As it regards the effects, in a general way, of these several kinds of baths, we may speak of them under two divisions, therapeutically considered. In the first, from the warm down to cold, we shall find a calming and soothing operation continued, with the reduced temperature of the water, to the most depressing sedative,—in fact, a reducing power; and in the second, from the upper degrees of warmth, a stimulating and strongly exciting operation. What a mischievous error, therefore, is the too common one of confounding *a warm* with a *hot bath*, and directing the one for the other, as if they were convertible terms expressing the same thing, instead of being in direct contrast with each other! It may serve to indicate the striking difference between the warm bath and the hot bath when I say that the first is a grateful hygienic agent, which almost everybody can make use of with benefit, in addition to its employment as a therapeutical one in the treatment of disease; whereas the hot bath is, or ought to be, a remedial agent to be used solely in disease, and even then with considerable caution and discernment."

SHOCCO SPRINGS are situated nine miles from Warrenton, in Warren County. They are a mild sulphurous saline water. My valued friend Dr. Howard, formerly of Warrenton, informs me that they are "mildly *aperient* and actively *diuretic*, producing, after a few days' use, free bilious evacuations, and that they are advantageously employed in the various diseases for which mild sulphur waters are usually prescribed."

Shocco is improved by a large hotel and comfortable cabins, that will pleasantly accommodate four hundred persons.

JONES' WHITE SULPHUR AND CHALYBEATE SPRINGS are located about five miles from Shocco, and eleven from Warrenton; they are improved for the accommodation of about three hundred and fifty visitors, and about that number may be found there at the height of the season.

The *White Sulphur* is a mild sulphurous saline water, and acts favorably in certain hepatic derangements, jaundice, dyspepsia, etc.

The *Chalybeate* is a strong ferruginous water; the *iron* is held in solution by carbonic acid. Dr. Howard considers it an excellent tonic, and "well suited for all those cases characterized by an enfeebled habit, and especially when the blood has been deprived of its normal proportion of iron. It displays marked efficacy in those whose blood has been robbed of this important element by malarious fevers, and in `chlorosis, amenorrhœa," etc.

KITTRELL'S SPRINGS.—Immediately on the railroad from Weldon to Raleigh, in the county of Granville, and half a mile from the village of Henderson, *Kittrell's Springs* are found. They have attracted public notice only for the last ten years, and as yet there is but little improvement at the place for the accommodation of visitors. The water of these springs has acquired considerable local reputation for the cure of various diseases, and particularly for scrofulous affections.

Chemical examinations have ascertained that the water holds in solution iron, magnesia, lime, alum, soda, and potassa.

These springs are probably destined to acquire a valuable medicinal reputation, and, when properly improved, to become a place of considerable valetudinary resort.

The WHITE SULPHUR SPRINGS, in *Catawba County*, are improved for the accommodation of a large number of visitors. They are delightfully situated, and are in a very salubrious and healthy climate.

In addition to the *sulphur waters*, there is here an excellent *chalybeate spring*, that has been long used to the great advantage of many invalids.

These springs can be conveniently reached by the distant visitant by making *Salisbury* a point in the travel from the north, south, or east.

CHAPTER XXIX.

SPRINGS OF SOUTH CAROLINA.

Glenn's—West's—Springs in Abbeville and Laurens Districts, etc.—Chick's—Williamstown Springs—Artesian Well in Charleston.

GLENN'S SPRINGS, in Spartansburg District, have considerable notoriety for their medicinal virtues.

Professor Shepard, of Charleston, states that the waters of these springs are strongly impregnated with sulphur, and that they also contain traces of sulphate of magnesia, with sulphate, percarbonate, and chloride of lime.

These springs are much resorted to by the people of the lower country. Their situation is pleasant, salubrious, and healthful, and their waters are highly esteemed by many, particularly in dyspeptic affections.

In the same district, and a few miles above the village of Spartansburg, there is a spring which is somewhat resorted to, and has acquired some local reputation.

WEST'S SPRING is in the neighborhood of *Glenn's*. It is a chalybeate of good promise.

Chalybeate springs are found in various parts of the State, particularly in Abbeville and Laurens Districts. In Laurens three or four chalybeate and sulphur fountains are known, that arise in the slate and hornblende formations that exist between the Ennoree and the Saluda, that are worthy of public attention.

I am indebted to the late Professor S. H. Dickson for the information that the springs most visited in South

Carolina are *Chick's Springs*, in Greenville District, on the Ennoree River, just below the mountains, and *Williamstown Springs*, between Anderson and Greenville.

CHICK'S SPRINGS are two in number. One is slightly sulphurous, and is used for hepatic and intestinal affections and cutaneous disorders. The other is a mild chalybeate, and is employed as a tonic.

The WILLIAMSTOWN SPRINGS have never been analyzed, so far as I know. They are supposed to be both tonic and alterative.

CHARLESTON ARTESIAN WELL.—The water obtained from this well has acquired some reputation as a remedial agent. An analysis of this water shows that one gallon contains nearly the third of an ounce of solid matter. Half of this is common salt, and three-quarters of the remainder are carbonate of soda. It has also traces of potash, bromide of magnesia, sulphate of lime, borate of soda, silica, and fluorine. It has been much used in Charleston, and many affirm that it relieves various derangements of the stomach and bowels. The late Professor Dickson informed me that horses are extremely fond of it, and that it is believed to act upon them beneficially, in promoting their ready fattening, and giving them a smooth and glossy coat. This water is exported in bottles and sold in considerable quantities in the Northern cities.

CHAPTER XXX.

SPRINGS OF GEORGIA.

Indian—Madison—Warm Springs—Gordon's—Catoosa Springs.

The INDIAN SPRINGS, in the county of Butts, are *sulphurous waters*, and are considerably visited and much relied upon as remedial agents. They have been used with excellent effect in chronic rheumatism, and for various diseases of the liver and stomach.

The MADISON SPRINGS have long been regarded as a pure and excellent *chalybeate*. They are found in the county of Madison, and are much visited by those who desire the use of iron tonics.

The WARM SPRINGS are in the county of Merriwether. Their temperature is 95°. They have acquired considerable reputation for the cure of *rheumatism, gout,* and other chronic affections for which such waters are commonly employed.

They are all situated in pleasant and salubrious districts, and so far elevated above the sea-board as to secure them against malarial influences.

Professor Richard D. Arnold, of Savannah, in a communication to Dr. Bell, thus speaks of this and the Indian Spring waters:—

"You have chalybeate springs in abundance at the North, but I doubt very much if any two springs can anywhere be found combining such decided medicinal qualities as the *Indian* and the *Merriwether Warm Springs.* They are also of very easy access from the

North. One of our fine sea-steamers would land a patient at our wharves in sixty hours from New York, and our railroad would convey him to within sixteen miles of the *Indian Springs* and about fifty of the *Warm Springs*. The former would be reached within four and a half days of travel from New York, and the latter within five and a half days."

GORDON'S SPRINGS, in the county of Murray, and ROWLAND'S SPRINGS, in the county of Cass, are *chalybeates*, and, within the last few years, are said to be attracting some attention from invalids.

CATOOSA SPRINGS are in the county of Catoosa, in the extreme western part of the State. They have not been analyzed, but are regarded as a saline chalybeate. They have been improved for the accommodation of several hundred persons, and are much visited during the watering-season.

CHAPTER XXXI.

SPRINGS OF ALABAMA.

Bladen Springs—Bailey's Spring—Tallahatta Springs.

ALABAMA has several springs of decidedly marked properties, the most noted of which is—

BLADEN SPRINGS, in the county of Clarke. These springs are within three miles of the Tombecbee River, eighty-five from Mobile, and seven from Coffeeville. The country surrounding them is broken and hilly, with a forest growth of pine, hickory, oak, etc., and is well supplied with wholesome water.

The accommodations at the springs are sufficient for several hundred visitors.

Six fountains, differing slightly from each other, issue from the earth within a small compass, furnishing an abundant supply of water.

Professor Brumby, of the University of Alabama, has analyzed the Bladen waters,* and from a wine pint obtained the following results:—

Sulphuretted hydrogen, quantity not ascertained.
Carbonic acid gas.................................... 4.075 cubic inches.
Chloride of sodium................................. 0.9625 "
Oxide of iron... 0.0300 "
Sulphate of lime..................................... 0.0019 "
Crenic aid.. 0.0912 "
Loss... 0.0400 "
Carbonate of soda.................................. 4.1112 "

---

* We are indebted to Dr. Bell's work on "Mineral and Thermal Springs" for many facts in reference to the springs of the extreme Southern States.

Carbonate of lime...............................	0.3437	cubic inches.
Carbonate of magnesia......................	0.1706	"
Silica of alumina.................................	0.2631	"
Apocrenic acid...................................	0.0750	"

The relatively large amount of carbonate of soda, with free carbonic acid, in this spring, classes it among the acidulous waters.

In various affections of the stomach, bowels, and kidneys, as well as in chronic rheumatism and diseases of the skin, the Bladen waters would prove valuable.

BAILEY'S SPRING is in Lauderdale County, nine miles from Florence, and fourteen from Tuscumbia. The water is cool, transparent, and essentially tasteless.

It has been chemically examined by Dr. Curry, of Knoxville, and is shown to contain sulphuretted hydrogen, carbonic acid, carbonates of soda and magnesia, oxide of iron in union with carbonic acid, carbonate of potash, and chloride of sodium.

The composition of this water shows that it would prove valuable in the various functional disorders of the abdominal and pelvic organs, in mercurial diseases, and generally in chronic affections of the skin, as well as in rheumatism and gout.

Besides the springs before noticed, the TALLAHATTA SPRINGS are much visited by persons in that part of the State. These waters are said to contain sulphur, magnesia, lime, and the salts of iron.

CHAPTER XXXII.

SPRINGS OF MISSISSIPPI.

Cooper's Well—Ocean Springs.

COOPER'S WELL is the most noted mineral fountain in Mississippi; it is in the county of Hinds, twelve miles west of Jackson, and four from Raymond, the shire town of the county, and near the Jackson Railroad.

The water rises in an artesian well, one hundred and seven feet deep, through solid sandstone rock. The surrounding country is broken and diversified, and is thought to be dry and salubrious. The water of this well is an active *saline chalybeate*.

An analysis of one gallon of the water, by Dr. J. Lawrence Smith, gives in gaseous contents:—

Oxygen	6.5 cubic inches.
Nitrogen	4.5 "
Carbonic acid	4.0 "

Solid contents:—

Sulphate of soda	11.705 grains.
Sulphate of magnesia	23.280 "
Sulphate of lime	32.132 "
Sulphate of potash	0.608 "
Sulphate of alumina	6.120 "
Chloride of sodium	8.360 "
Chloride of calcium	4.322 "
Chloride of magnesium	3.480 "
Peroxide of iron	3.362 "
Crenate of lime	0.311 "
Crenate of silica	1.801 "
	105.471 "

The *deposit* obtained by evaporating the water contains in one hundred and five grains—

Water...	38 grains.
Chloride of lime...	2 "
Sulphate of lime...	25 "
Peroxide of iron...	25 "

This water is said to lose none of its qualities by being kept from the fountain.

The water of Cooper's Well enjoys a high reputation in dyspepsia and the various intestinal diseases of long standing; in liver complaints, chronic inflammation of the bladder, in dropsy, and especially in *chronic diarrhœa*. Its analysis shows that it is a medicinal agent of very decided powers.

Dr. Foster's case, as reported by Dr. I. M. Sims, of Montgomery, Alabama, is very remarkable. It was a chronic diarrhœa in its worst form, emaciation extreme, skin dry, eyes sunken, expression so ghastly as to cause a lady to faint at sight of him, small and feeble pulse, frequent and copious dejections from the bowels. Dr. F. commenced by taking a wineglassful of the water four times during the day, gradually increasing the amount until he drank a pint in the course of the day. In eight weeks he was cured, and returned home a well man.

The medical properties of this water are *cathartic* or *aperient*, according to the quantity taken. It also exerts diuretic, sudorific, tonic, and alterative influences upon the system. As an alterative, its influence upon the blood and upon diseased organs and tissues is especially worthy of notice. The efficacy of the water in various diseases usually unmanageable in the hands of physicians commends it to the attention of the medical profession; while the promptness and certainty of its action entitle it to the hopeful consideration of the invalid.

To the various diseases of the abdominal and pelvic

regions this water is well adapted. Among these, diseases of the biliary organs unattended with obstinate obstructions, dyspeptic depravities, and chronic diarrhœa, are treated by it with great success.

While as a remedy in that scourge of the South, chronic diarrhœa, this water may be looked to generally with great hope, a careful discrimination is nevertheless necessary in using it in such cases, for, if the diarrhœa be connected with, or dependent upon, a diseased condition of the lungs, it would prove positively injurious, and hasten a fatal tendency.

The OCEAN SPRINGS are situated in the pine hills of Jackson County, five miles from the town of Biloxi, half a mile from Biloxi Bay, and near Fort Bayou.

One gallon of this water has in gaseous contents—

Carbonic acid	4.632 grains.
Sulphuretted hydrogen	0.481 "

In solid contents —

Chloride of sodium	47.770 grains.
Chloride of calcium	3.882 "
Chloride of magnesia	4.989 "
Protoxide of iron	4.712 "

With *traces* of iodine, organic matter, chloride of potassium, and alumina.

Dr. Bell, in quoting Dr. J. Lawrence Smith, remarks that the iron is doubtless in combination with both the sulphuretted hydrogen and carbonic acid gases; the excess of carbonic acid holding both these combinations in solution.

Dr. Austin, of New Orleans, in a letter to Dr. Bell, states that striking cures have been wrought by these waters in many chronic diseases; among them are affections of the skin, scrofula, dyspepsia, and strumous ophthalmia.

The Ocean Springs are very easy of approach both from New Orleans and Mobile, being about ninety miles distant from both places.

CHAPTER XXXIII.

SPRINGS OF ARKANSAS AND FLORIDA.

THE HOT SPRINGS OF ARKANSAS, commonly known as the *Washita Springs,* are among the most remarkable thermal fountains in the world.

They are located in Hot Springs County, latitude 34° 5', longitude 16° 1', about fifty-five miles southwest from Little Rock. Hot Spring Valley runs due north and south between the two spurs of the Ozark Mountains, through which a bold creek heads its way over an almost unbroken bed of slate, emptying into the Ouachita River, about five miles distant.

Hot Springs Mountain lies on the east of the valley, from the west side of which gush the Hot Springs, rising upwards of two hundred feet from the level of the valley, and from the very base, and many from the bottom of the creek; the valley is about three hundred feet wide, and eight hundred yards in length. Fifty-four hot springs have been tested in temperature, whilst many at the bottom of the creek, and under the ledges, cannot be, except with too great labor. About 350 gallons of hot water are discharged into the creek per minute from said fifty-four springs, which affords the enormous yield of 504,000 gallons in twenty-four hours. The largest spring discharges 60 gallons of hot water per minute, at a temperature of 148°, and will cook eggs in fifteen minutes. There is only one hot spring on the west side of the creek, called the alum, and immediately opposite, on the east side, one of sulphur, though very slightly impregnated with either. There are only four cold-water springs in the vicinity of the Hot Springs, viz.: one chalybeate 70° temperature, south end of valley, two freestone 70° temperature, north end, and one chalybeate 69°, quarter-mile northeast. There are two wells in the valley about twenty feet deep, 70° temperature. Water boils on the summit of Hot Spring Mountain at 208°: scant five hundred and twenty feet elevation for each degree less 212°, gives nearly twenty-one hundred feet above the level of the sea. In Hot Springs Valley water boils at 209°, which makes Hot Springs Mountain five hundred and sixty feet above the valley.

On the summit of the mountain are heavy pine and oak timber, abounding with clusters of grape-vines, huge masses of quartz rock, apparently upheaved by some convulsion of nature; immediately below the summit, sharp-cornered broken honey-comb rocks, with sparkling surfaces; still lower, a heavy undergrowth of pines and other trees, and from thence, where the Hot Springs flow to the base, calcareous tufa.

These springs vary in temperature from 100° to 148° Fahr. These results were arrived at by testing them at three different hours of the day, viz., between four and six o'clock A.M., at twelve M., and between four and six P.M. There is no perceptible difference in the temperature tested at these several periods.

The *vapor baths* that have been constructed here stand at 112°, the *douche*, a spirit bath, at 120°, and the *saving bath* at 116°, the two latter varying slightly, from the negligence of the attendants.

The *analysis*, by Dr. Owen, of what is termed the *Rector House Well*, shows it to contain bicarbonates of lime, magnesia, and iron, subcarbonates of magnesia, iron, and soda, chloride of sodium, and sulphates of soda and magnesia in small quantities.

The medicinal effect of this water, internally used, is slightly *aperient*, *antacid*, and *tonic*.

It has been observed by Dr. Owen that all the springs,

wells, and water-courses of this region of country partake of some mineral impregnation in a greater or less degree.

A heavy fog continually hangs over these springs, and upon the sides of the mountains, giving the neighborhood the appearance, at a little distance, of a number of furnaces in active operation.

The water is, essentially, tasteless, very clear, pure, and transparent, and does not deposit sediment by standing.

Near the edges of the springs is found luxuriously growing a species of green *algæ*, which seems to delight in these natural hotbeds, while the sides of the mountain are covered with luxuriant vines, continually watered by the condensation of the vapor from the springs.

Mr. Featherstonehaugh, in his "Geological Report of 1835," remarks that the lofty ridges around these springs consist of old red sandstone formation. Upon the eastern ridge are found fragments of the rock, often ferruginous, with conglomerate united by ferruginous cement. Upon the side of this ridge is found *travertin*, deposited by the mineral waters, extending the distance of one hundred and fifty yards, resting upon the old red sandstone, presenting, sometimes, abrupt escarpments of from fifteen to twenty feet.

Dr. G. W. Lawrence, a gentleman eminent in the profession, and who for many years has resided as a practitioner at these springs, has kindly favored me with a communication upon their therapeutic character, from which I make the following extracts:—

"As a stimulant, when taken internally, it arouses the *absorbent* and *secreting system, stimulates the hæmic glands*, produces more rapid *metamorphosis*, and '*alterant*' action is the result. The water is easily assimilated and brought rapidly into the circulating system; thus producing, when elaborated, active *eliminative* agency. Thus we have all the blood-making organs aroused by the pure, tasteless, inodorous, natural stimulant, through the medium of the blood. It rapidly courses every part of the circulation, and if no organic disease exists, the efficacy, *as an adjunct*, in the treatment of *all blood diseases*, is sometimes truly marvelous.

"In uterine diseases, as a class, these waters are unrivaled in efficacy. In that tedious form of *chronic metritis* where ulcerative action ensues, and *neuralgia* and *functional* difficulties follow, no agency can be made more valuable to the sufferer.

"Where *sterility is alone functional*, the causes can generally be relieved by the judicious use (internally and externally) of the waters. Cutaneous diseases, the opprobrium generally of the medical profession, especially when of a *specific type*, are treated here with the greatest advantages,—not only from the agreeable detergent action of the baths, or the maceration of old morbid surface-tissues that are cleansed, but in the treatment of all skin-diseases, where we find *integumentary alterations or lesions* existing, the natural tepid, warm, and hot baths in efficiency cannot be excelled. In all *rheumatic conditions of the system*, after the acute or inflammatory action subsides, the thermal waters enjoy great celebrity for their good qualities and curative properties. In the treatment of *gout and gouty rheumatism*, the waters have like reputation in controlling the '*diathesis*,' if persistently used as directed. As remedial *adjuncts* in the treatment of *scrofula, syphilis, mercurio-syphilis, mercurial diseases*, and *climatic (malarial) ills*, where prompt '*depurative*' and '*eliminative*' agency is demanded, these waters have no superior, in fact stand unrivaled, in combined properties, for that agency. In all diseases of the brain or lesions of the spinal marrow, these waters are *positively* injurious. Experience, with careful circumspection, satisfies me that the waters should not be used in *epilepsy*, except it is purely of *functional* origin. Females should avoid, if possible, the treatment of chronic diseases *during pregnancy*, as unpleasant results are very apt to follow general bathing.

"In all diseases of the lungs, or bronchial tubes, *without specific origin*, all natural thermal waters are undesirable, as they oppress respiration by stimulating circulatory action, and cause an afflux of blood to the bronchial surfaces. In organic diseases of the heart, thermal waters (either natural or artificial) should not be used."

About three miles from the Hot Springs there is a *chalybeate spring*, which is said to be of very fine quality.

In Montgomery County, forty miles from the Hot Springs, is a spring known as "*Bill Iron's Salt Sulphur*," which is said to possess highly exhilarating properties, so much so as to produce the peculiar symptoms of incipient intoxication.

SPRINGS OF FLORIDA.

There are light sulphurous waters in various parts of Florida, but none have become places of large visitation. Among these may be mentioned the *Sulphur Spring* near Tampa. It arises from a bed of limestone. The water is remarkably clear and transparent, and forms a basin at its source eighteen feet deep.

There are several springs on the St. John's and Suwanee Rivers, known as the *Magnolia*, the *Walake*, and the *Enterprise* Springs,—all sulphurous.

At the Magnolia, a *sanatarium* has been established for the reception of invalids who may wish to spend the winter in that climate.

We are told, by a writer in the *Floridian Journal*, that Florida greatly *abounds in mineral waters*, and that their solid contents consist generally of the sulphates of lime, soda, and magnesia, with oxide of iron; their gaseous contents of sulphuretted hydrogen, carbonic acid, and nitrogen gases. But too little, as yet, is known of these springs to determine with certainty their relative or positive merits.

CHAPTER XXXIV.

MINERAL SPRINGS OF NEW YORK.

Saratoga and Ballston Group—Classification of Waters—Geological Position—Thermalization of Waters—Analysis of Various Springs, etc.

NEXT to Virginia, New York is more distinguished for the number and variety of her mineral springs than any State of the Union. With less variety in the composition of her waters than Virginia, she nevertheless possesses some of very high medicinal character, and that have more largely attracted public attention than any other waters in America. I allude, of course, to the distinguished group known as the *Saratoga and Ballston Springs.* This entire group possesses essentially the same properties and virtues; the difference between the several springs consisting merely in the proportions of their relative gaseous and saline contents.

The famous series of springs at Saratoga comprise the several springs known as *Congress, Putnam, Pavilion, High Rock, Iodine, Flat Rock, Hamilton, Columbian, Washington, Empire, Saratoga Alum, Geyser, Star, Halthorn, Excelsior, Seltzer,* and *Red Spring.*

The village of Ballston Spa lies about seven miles southwest from Saratoga. The large resort to this place, on account of its mineral springs, makes it, like Saratoga, a place of considerable notoriety.

The mineral springs of *Ballston* comprise the *Sans Souci, Low's Park,* the *New* and the *Old Washington Springs,* and the *Sulphur Spring.*

The waters of Ballston, with the exception of the

Sulphur Spring, evidently belong to the same class with those of the Saratoga group. And although they do not contain quite so large a proportion of saline qualities as some of the Saratoga fountains, they are, nevertheless, entitled to rank high among the acidulous chalybeate waters of our country.

In classifying the Saratoga and Ballston springs, we may well regard them as *acidulo-saline* or *carbonated saline waters*. Their large amount of carbonic acid gas and of carbonates, with their heavy impregnation with chloride of sodium, distinctly assigns them to this class.

The great Appalachian chain of geological upheavings, extending through Virginia and West Virginia, and furnishing such an extensive series of thermal and medicated waters, is probably on the same or a parallel axis with that which gives the famous waters of Saratoga and Ballston.

The fact that the various springs of Saratoga and Ballston hold in solution essentially the same ingredients, and differ from each other only in the quantity of ingredients common to all, goes to show that they derive their distinctive qualities from one common source, but are modified to some extent in their passage to the surface of the earth by the peculiar character of the different strata through which they have passed.

"If," says Dr. Bell,* " we admit the correctness of Dr. Daubeny's observation, that the temperature of the water of the Congress spring at Saratoga, 51° Fahr., is three or four degrees above the mean temperature of the earth at this place, we can give credence to the opinion of the thermal origin of the water, and of the mode of extrication of the carbonic acid so largely found; it being brought about by subterranean heat acting on limestone rocks. The first process would consist of the junction of carbonic acid coming through

---

* Mineral and Thermal Waters of the United States and Canada.

the clefts and small canals, with the meteoric water which had reached its greatest depth and was beginning to rise in larger canals. The second process would be the decomposition and solution of portions of certain rocks, and the formation of acidulous springs, rich in carbonic acid and carbonates. The same heat which would drive off carbonic acid from limestone would readily raise the temperature of the meteoric water which finds its way into the interior of the earth, and we should then have thermal—warm and hot—springs. Reasoning in this way, we can easily adopt the views of those who maintain that carbonated and thermal springs are similar in their mineral, and still more in their geological, position, and seem to be plainly referable to the same system of causes."

Admitting the correctness of the supposition that subterranean heat may be sufficient to eliminate carbonic acid from limestone, and so to heat meteoric water in the bowels of the earth as to return it to the surface in the form of hot and warm springs, a question for the curious still remains to be mooted. Is this subterranean heat volcanic, and consequently local, or is it from the great "central heat" of the earth, contended for by Mr. Daubeny and others? Many geological appearances in the regions in which we find thermal waters, not to mention the extensive upheavings and displacement of strata generally found in the neighborhood of such springs, lend some countenance to the volcanic origin of such waters. On the other hand, the theory of the central heat of the earth, which alleges that the earth's heat increases about one degree for every hundred feet we descend in it, while it has been occasionally sustained by deep artesian borings, has, on the other hand, been so often refuted by such borings, that it seems unsafe, in the absence of more conclusive proof, to adopt it as a fixed and well-determined fact.

CONGRESS SPRING.

The following is the analysis of the Congress water, as made by Dr. Steel:—

He states that the temperature of the water is 50°. Dr. Daubeny marks it at 51° Fahr.

Both its temperature and quantity are said to be the same at all seasons.

One gallon of the water yields—

Chloride of sodium...	385.0
Hydriodate of soda...	3.5
Carbonate of soda..	
Bicarbonate of soda..	8.982
Carbonate of magnesia...	
Bicarbonate of magnesia..	95.788
Carbonate of lime...	98.098
Carbonate of iron...	5.075
Silica	1.5
Hydrobromate of potassa...	a trace.
	597.943

Gaseous contents:—

Carbonic acid...................................	311	cubic inches.
Atmospheric air.................................	7	" "
	318	" "

Dr. Chilton's examination of this water, as given by Dr. North, differs somewhat from the above. He found a minute portion of alum, sulphate of soda, iodide of sodium, and bromide of potassium, to the amount of 5.920 grains to the gallon of water. According to his estimates, the solid and gaseous contents of the water in one gallon are as follows:—

Solid contents...................................	543.998	grains.
Carbonic acid...................................	284.65	cubic inches.
Atmospheric air.................................	5.41	" "
	290.06	" "

Iodine was first discovered in these waters in 1828, and was announced in the *American Journal of Science* in 1829. In 1830, Mr. A. A. Hays detected bromine and potash in the water. The quantity of these ingredients is, however, very small, and to detect them with certainty it is necessary to operate on a large quantity.

PUTNAM SPRING.—This spring, bearing the name of its proprietor, is regarded as the richest chalybeate in the Saratoga group. It is reported as containing seven grains of the carbonate of iron to the gallon, in addition to the salts common to it and the other springs. This, comparatively, is a heavy chalybeate impregnation. The famous *Pyrmont Spring*, in Westphalia, which enjoys, perhaps, the largest European reputation as an iron tonic, contains, agreeably to the analysis of M. Westrum, but $8\frac{1}{2}$ grains of iron to the gallon; while the celebrated *Pouhon*, at Spa, in Belgium, little if any less distinguished as a chalybeate tonic, contains but 5.24 grains of iron to the gallon, according to the analysis of the celebrated Bergmann.

PAVILION SPRING.—The saline contents of the water of this spring are less than those of the Congress, being 311.71 grains in the gallon. It, however, exceeds the latter in the proportion of its carbonic acid, of which it has 359.05 cubic inches to the gallon. This spring is near the Columbian Hotel, and not far from the centre of the town.

UNION SPRING.—By Dr. Chilton's analysis, the water of this spring is shown to contain 392.907 grains of solid contents in the gallon. Its amount of carbonic acid is somewhat less than is found in the Pavilion, being 344.16 cubic inches in the gallon of water. This spring is in the eastern part of the town, and not far from the road leading to Schuylerville.

HIGH ROCK.—This spring, with its conical inclosure of *calcareous tufa*, evidently the deposit of its own waters, deserves to be regarded among the interesting curiosities of our country. The venerable Dr. Seaman remarks, in reference to it, that if it "had been upon the borders of the Lago d'Agnano, the noted *Grotto del Cane*, which burdens almost every book which treats upon the carbonic acid gas since the peculiar properties of that air have been known, would never have been heard of beyond the environs of Naples, while this fountain, in its place, would have been deservedly celebrated in story, and spread upon canvas, to the admiration of the world, as one of its greatest curiosities."

This unique conical structure is composed of the carbonates of lime and magnesia, with the oxide of iron, and a portion of sand and clay. When broken, it exhibits the impression of leaves and twigs of trees. Its circumference at its base is about twenty-six feet, and its perpendicular height four feet; from the top of the rock to the surface of the water, two feet; depth of water in the cavity of the rock, about seven and a half feet. The hole at the top of the rock through which the water is dipped is circular, and measures about ten inches across.

As early as 1767, this spring was visited by Mr. Wm. Johnson, who used its waters with benefit for gout, and from this period it came rapidly into the notice and regard of the colonists. In the years 1784 and 1785 some accommodations were constructed for invalids, and about this period the springs known as Flat Rock, the President, and Red Spring, first attracted attention.

Dr. Steel, to whose "Analysis" I am indebted for this history, remarks that "the extravagant stories told by the first settlers of the astonishing effects of the 'High Rock' waters, in the cure of almost every species of disease, are still remembered and repeated by their too credulous descendants. This, in connection with the singular and mysterious character of the rock, continued to attach an importance to the waters, in the eyes of the vulgar, to which no other fountain will ever attain."

The temperature of the High Rock water is 48°; its specific gravity, 1006.85, when the barometer stood at 29.05 inches—pure water being 1000. Analysis shows that it contains 345.68 grains of solid ingredients, and 309 cubic inches of gaseous contents, to the gallon of water. Each gallon holds in solution 5.58 grains of carbonate of iron.

The IODINE, or, as it is sometimes called, *Walton Spring*, contains, according to the examinations of Professor Emmons, 3.5 grains of hydriodate of soda to the gallon of water. Its saline ingredients do not differ essentially from those of the neighboring fountains. Its chalybeate impregnation is somewhat greater than the water of the Congress Spring, but less than that of the Putnam, Union, Pavilion, and others.

Its temperature is rendered at 47° Fahrenheit.

The *Flat Rock*, *Hamilton*, *Columbian*, and *Washington Springs*, of which Dr. Steel gives the analysis, very nearly resemble each other, and those already described, in their general saline and gaseous character. The *Flat Rock* contains 5.39 grains of the carbonate of iron to the gallon, the *Hamilton* 5.39, the *Columbian* 5.58, and the *Washington* 3.25.

EMPIRE SPRING.—This fountain is now attracting considerable attention. The relatively larger portion of *iodine*, and smaller portions of iron and earthy salts, contained in this water, in comparison with its neighboring springs, suggest to the medical mind a preference for it in the treatment of several formidable chronic affections.

The following is Professor Emmons's analysis of one gallon of the water:—

Chloride of sodium	269.696
Bicarbonate of lime	141.824
Bicarbonate of magnesia	41.984
Bicarbonate of soda	30.848
Hydriodate of soda or iodine	12.000
Bicarbonate of iron	a trace.
	496.352

Specific gravity 1039.

SARATOGA ALUM.—This is one of the Saratoga group of comparatively recent development. Its analysis by Dr. J. G. Pohle, of New York, which follows, is calculated to give it a high position among its most distinguished compeers:—

Chloride of sodium	565.300
Chloride of potassium	.357
Chlorides of calcium and magnesia	traces.
Bicarbonate of soda	6.752
Bicarbonate of lime	56.852
Bicarbonate of magnesia	20.480
Bicarbonate of iron	1.724
Sulphate of lime	.448
Sulphate of magnesia	.288
Sulphate of soda	2.500
Sulphate of potassa	.370
Silicic acid	1.460
Alumina	.380
Per gallon	656.911

Free carbonic acid gas	212 cubic inches.
Atmospheric air	4 " "
Per gallon	216 " "

It will be observed from this analysis that this water is about ten per cent. greater in mineral properties than the celebrated Congress Spring; while it is four times that of Baden-Baden in Austria, twice that of Vichy in France, nearly three times greater than the renowned Seltzer of Germany, and five times greater than that of Aix-la-Chapelle in Prussia.

The *Geyser*, or "*Spouting Spring*," on the Ballston road, one and a half miles south of the principal hotels at Saratoga, is very remarkable for the amount of its constituent ingredients, both solid and gaseous. Prof. Chandler represents it as containing 991.546 grains of solid matters to the gallon, with 454.082 cubic inches of carbonic acid gas.

The STAR, HALTHORN, EXCELSIOR, SELTZER, and RED SPRING all resemble, in the general character of their waters, the springs of the famous Saratoga group just described.

BALLSTON SPRINGS.

The village of Ballston is situated seven miles southwest from Saratoga. It derives its name from the late C. Eliphalet Ball, who with a number of his congregation settled near the site of the village at the time the springs were first discovered. These mineral springs are situated in a deep marshy valley, through which passes a branch of the Kayaderosseras Creek. They were discovered in 1769.

Of the springs composing the Ballston group of acidulous chalybeate waters, the following may be mentioned: the *Sans Souci, Park, Low's Well,* the *United States, Franklin,* and *Fulton Chalybeate.* Dr. Steel remarks that these waters evidently belong to the same class with those at Saratoga; and if they do not contain so large a portion of the saline properties as some of the fountains at the latter place, which is very manifest, both from the taste and the effects, they are, unquestionably, entitled to rank among the best acidulous chalybeate waters which this or any other country affords.

In addition to the acidulous saline chalybeate waters of Ballston Spa, there are several sulphurous springs in the neighborhood, not regarded, however, as very strong, which probably owe their peculiar character to the decomposition of the sulphuret of iron which abounds in the argillaceous slate formation common to this region.

Sans Souci Spring contains, by analysis, in one gallon of its water—

Chloride of sodium	143.733 grains.
Bicarbonate of soda	12.66 "
Bicarbonate of magnesia	39.01 "
Carbonate of lime	43.407 "
Carbonate of iron	5.95 "
Hydriodate of soda	1.3 "
Silex	1. "
	247.15 "

The waters of *Low's Well* are regarded as being almost identical with those of the Sans Souci.

In the waters of the *Park Well* Dr. Steel demonstrated the existence of $6\frac{1}{2}$ grains of the carbonate of iron in a gallon of the water; a somewhat larger quantity than is found in any of the other fountains.

The *United States Spring*, according to Dr. Beck's analysis, contains in one pint of the water—

Chloride of sodium	53.12 grains.
Carbonate of soda	2.11 "
Carbonate of magnesia	0.72 "
Carbonate of lime, with a little oxide of iron	3.65 "
Sulphate of soda	0.22 "
Silica	1.00 "
	60.82 "

Carbonic acid, 30.50.
Temperature, 50° F., which does not vary through the year.

It will be seen, by comparing the analysis of this with the Congress Spring, that the latter contains a much larger amount, both of solid and gaseous contents, than the former.

The *Franklin Mineral Spring* has been analyzed by Prof. C. F. Chandler, with the following results:—

One U.S. gallon, 231 cubic inches, contains—

Chloride of sodium	659.344 grains.
Chloride of potassium	33.930 "
Bromide of sodium	4.665 "
Iodide of sodium	.235 "
Fluoride of calcium	trace.
Bicarbonate of lithia	6.787 "
Bicarbonate of soda	94.604 "
Bicarbonate of magnesia	177.868 "
Bicarbonate of lime	202.332 "
Bicarbonate of strontia	.002 "
Bicarbonate of baryta	1.231 "
Bicarbonate of iron	1.609 "
Sulphate of potassa	.762 "
Phosphate of soda	.011 "
Biborate of soda	trace.
Alumina	.263 "
Silica	.735 "
Organic matter	trace.
Total	1184.368 "
Carbonic acid gas	460.066 cubic inches.
Density	1.0115 "
Temperature	52°

CHAPTER XXXV.

NEW YORK MINERAL WATERS—CONTINUED.

Improper Use of the Saratoga Waters, and its Evils—Injurious Advice and Errors of Opinion as to the Nature and Use of Mineral Waters—Diseases for which the Saratoga Waters may be prescribed—Albany Artesian Mineral Well—Reed's Mineral Spring—Halleck's Spring, etc.

It is well remarked by Dr. Steel, long the resident physician at Saratoga,* that "these waters are so generally used, and their effects so seldom injurious, particularly to persons in health, that almost every one who has ever drank of them assumes the prerogative of directing their use to others; and were these directions always the result of experience and observation, they certainly would be less objectionable; but there are numerous persons that flock about the springs during the drinking season without any knowledge of the composition of the waters, and little or none of their effects, who continue to dispose of their directions to the ignorant and unwary with no other effect than to injure the reputation of the water and destroy the prospects of the diseased.

"Many persons who resort to the springs for the restoration of health seem to be governed by the idea that they are to recover in proportion to the amount they drink; and although many who are in health may, and frequently do, swallow down enormous amounts of the water with apparent impunity, it does not fol-

---

* Analysis of the Mineral Waters of Saratoga and Ballston.

low that those whose stomachs are enfeebled by disease can take the same quantity with the same effect. Stomachs of this description frequently reject large portions of the water, and thereby protect the system from the disastrous consequences that would otherwise follow. But when it happens to be retained, the result is indeed distressing. The pulse becomes quick and feeble, the extremities cold, the head painful and dizzy, the bowels swollen and tender, and the whole train of nervous affections alarmingly increased; and should the unfortunate sufferer survive the effects of his imprudence, it is only to a renewal of his worst apprehensions, from a loss of confidence in what he most probably considered a last resort."

The above sensible remarks of a gentleman long accustomed to prescribing mineral waters, and entirely familiar with their potent influences for good when properly used, or for evil when improperly employed, commend themselves with great force to invalids generally who resort to mineral fountains for relief.

The injury done to invalids at mineral springs by hasty and well-intentioned but ignorant and injudicious advice, both as to the applicabilities of the waters and the method of using them, by persons they may chance to meet, can scarcely be overrated. Various instances have occurred of invalids being speedily destroyed by improperly using mineral waters, under the injudicious advice of ignorant and officious persons, and still more frequently have diseases been aggravated and confirmed through such reprehensible officiousness, that might have been cured under sensible and judicious instructions. Besides, the idea that is often spontaneously in the mind of the invalid, that it is "only water" he is drinking, and that it can do no harm if it does no good, is simply an imposition on his own good sense, and upon the feeblest powers of ratiocination. These impressions upon the mind, vague though they may be, are nevertheless occasionally sufficiently strong to control the action. Such views are most apt to find a lodgment in the minds of those who have decided to altogether repudiate medicine, commonly so called, and to seek their lost health by the use of mineral waters, not remembering that mineral waters are *medicines*, and could be of no service if they were not. Under the false impression of their non-medicinal nature, such persons will often take into their stomachs, in the form of draught after draught of sulphur waters, more medicinal material in one day than a judicious physician would give them in pill or potion in an entire week.

It was such persistent abuse of mineral waters on the Continent that induced Henry IV. of France to decree a royal edict that no person should enter upon the use of a mineral water in his dominion until his case had been professionally examined and the suitableness and manner of using the water prescribed.

When Americans shall have acquired more prudence upon this subject, and learned to inquire more carefully into the adaptedness of mineral waters to their diseases, before committing themselves to their use, far more good will be derived by the invalid; our mineral waters will be appreciated, and their character better established in public confidence.

DISEASES FOR WHICH THE SARATOGA WATERS MAY BE PRESCRIBED.

In reference to the proper manner of using the Saratoga waters, as well as to the diseases for which they may be prescribed, I shall confine my remarks to a few *general observations* having reference to the usual proper use of such agents, knowing that *particular directions for the* individual case can be most prudently and safely given to the patient by experienced practitioners resident at the springs, and after such careful personal investigation of the case, and with such discriminating views of its pathology, as personal examination can alone, in most cases, satisfactorily determine.

The entire group of the Saratoga and Ballston waters may properly be regarded, as I have before stated, as distinctly belonging to the saline acidulous class, with chalybeate salts so prominent in some of them as to modify, in an important degree, their influence upon the animal economy. Their prominent therapeutic effects are those of active aperient and diuretic action.

A numerous class of visitors at mineral springs are those who are rather threatened with, than actually laboring under, a distinctly located disease. As prominent in this class of visitors, we find those who suffer under a preternatural fullness of the blood-vessels, and especially of the veins, with a tendency to congestion in some of the large internal organs, with a sense of fullness or heaviness in the abdominal regions. This condition is often occasioned from slow and imperfect digestion, and, consequently, by too long retention of food in the stomach, from local and general accumulations in the large intestines, and not uncommonly from an engorged liver or spleen, with a sluggish circulation, and sometimes a throbbing sensation in the portal system. This morbid state of the system is made to bear different names as one or another organ seems to be more especially affected.

The morbid tendencies of this condition are very numerous. Even in its incipiency it is prone, from hygienic or morbid causes, to run into obstinate congestions, irritations, or actual inflammations. Sometimes it results in cephalic or pectoral accumulations, giving occasion for apoplexy, asthma, etc. In other cases, the system seems to make a violent external effort to relieve its internal oppressions through an acute attack of rheumatism or gout; or by eruptions upon the surface, carbuncles, boils, or erysipelatous inflammations. The most common winding up of this general plethoric condition is a confirmed dyspepsia, attended with faulty and irregular secretions from the liver, ultimately giving rise to intestinal or thoracic neuralgia.

Space will not allow me to trace out the various and multiform disorders and disorganizations that may, and often do, result from the venous plethora and abdominal accumulations alluded to; nor is this, perhaps, the proper place to do so. I remark, however, that, in the condition of the system alluded to, and especially in its early stages, the Saratoga waters, and of choice the more purgative of them, afford a remedy entitled to great confidence, and, generally, speedily beneficial in its effects.

In such cases they should be so used as to produce copious evacuations from the bowels for two or three weeks. The more purgative waters, such as the Congress Spring, being taken early in the morning to produce this effect, the patient may, with advantage, use small quantities of some of the more ferruginated waters in the evening, such as the Putnam, or High Rock Spring.

In recent attacks of biliary affections, unattended with fever or general excitement, the Congress waters have proved very beneficial. In such cases, Dr. Steel, long a resident physician at the springs, says he was in the habit of giving a few grains of calomel or blue pill at night, and following it in the morning with a sufficient quantity of water to move the bowels briskly two or three times. A few doses of this description usually put the bowels in a situation to be more easily acted upon by the water alone. In the *more advanced stage of bilious affections,* says Dr. Steel, "where the organization of the liver and other viscera has materially suffered, and the disposition to general *hydrops,* indicated by the enlargement of the extremities, fullness of the abdomen, etc., the waters are, all of them, manifestly injurious, and ought not to be admitted, even as an adjunctive remedy."

In the various *dyspeptic depravities* these waters have long maintained a high and well-deserved reputation. The Congress Spring is most generally used for these affections. It is best taken in the morning for such cases, about an hour before breakfast, in sufficient quantity to move the bowels gently once or twice. For this purpose, from two to four or five tumblerfuls, taken at intervals of ten or fifteen minutes apart, will generally be sufficient.

In *calculous* or *nephritic complaints*, these waters have been long employed with great advantage, and well-attested instances are given of their effecting complete cures in such cases. The water, in such diseases, should be so drunk as to keep the bowels gently open and to keep up an increased secretion from the kidneys. In such cases, the use of the warm bath is an important auxiliary. Its temperature should be about 100° Fahr., and the patient remain in it from thirty to sixty minutes.

In *chronic rheumatism*, Dr. Steel asserts that the waters have been long employed with advantage. In such cases, he gives preference to the Congress Spring.

For *arthritis* or *gout*, the waters are regarded as an uncertain remedy. In the early or forming stages of the disorder they may prove beneficial, but when the disease has become confirmed, and is of long continuance, the effects of the water are doubtful, and cases have occurred where their use induced a recurrence of the paroxysm.

In *ill-conditioned ulcers and cutaneous eruptions*, as well as in the enfeebled condition of the system following a *protracted mercurial course*, the use of the waters has proved very beneficial.

Scrofula is another disease in which the Saratoga waters have been often used, and Dr. Steel remarks that "experience abundantly sanctions the belief of their utility in that complaint."

The large proportion of iodine which Professor Emmons detects in the Empire Spring seems clearly to indicate a preference for that fountain in the treatment of this class of affections.

In *dropsical affections* the Saratoga waters should only be prescribed under careful discrimination. When the disease depends upon long-continued organic derangement, they will prove injurious. On the other hand, when the affection is recent, and dependent upon the want of sufficient action in the absorbent vessels, they will be beneficial, and their use in such cases will probably result in removing the morbid accumulations.

Paralysis, under the active purgative operation of the waters, is sometimes benefited.

Chlorosis and other complaints peculiar to females are often treated by these waters with good success. In such cases, the waters in which the *tonic* properties most abound are to be preferred, and much advantage will generally be derived from frequent bathing, and pleasurable exercise unconnected with exhaustion or fatigue.

In *phthisical complaints* that arise from a primary affection of the lungs, the Saratoga waters are injurious, and ought not to be used. But in congestions of the bronchial surfaces, as well as in translated or sympathetic affections from abdominal origin making a lodgment in the chest, and unattended with any general strumous tendency, the waters of the Empire Spring might, probably, be safely and advantageously employed.

ALBANY ARTESIAN MINERAL WELLS.—Messrs. Boyd and McCullock, in boring for pure water to supply their brewery, struck at the depth of four hundred and eighty feet a saline water abounding in the carbonates and carbonic acid, and emitting at the same time carburetted hydrogen or burning gas. On continuing the boring to the depth of six hundred feet, the flow of the carbonated water and gas continued. Another boring was effected to the same depth, a few rods from the first, with the same results and the singular addition of the escape of sulphuretted hydrogen gas from a small stream of water that was struck at thirty feet below the surface. From this, Dr. Beck concludes that "in the same slate formation, though at different depths, sulphuretted hydrogen, carburetted hydrogen, and carbonic acid gases are abundantly evolved." The same writer thinks it probable that carbonated waters might be found by boring at any point on the range from Saratoga to Albany.

The temperature of the water of the Albany well is 51° to 52° Fahr. Its specific gravity is 1.00900.

Dr. Beck's analysis of one pint of water shows the following results:—

Chloride of sodium...	59.00 grains.
Carbonate of soda...	5.00 "
Carbonate of lime...	4.00 "
Carbonate of magnesia.......................................	1.50 "
Carbonate of iron, with a little silica.....................	1.00 "
Chloride of calcium...	0.50 "
	71.00 "

Gaseous contents, 28 cubic inches.

REED'S MINERAL SPRING, in Washington County, is an acidulous spring, not very dissimilar from the waters of Saratoga, but containing less gas, and consequently less sparkling. Its taste is somewhat acidulous.

HALLECK'S SPRING, in Oneida County, and near the village of Hampton, was discovered by boring to the depth of one hundred and six feet into a solid rock.

Professor Noyes analyzed this water, and obtained from one pint the following results:—

Chloride of sodium...	78.00 grains.
Chloride of calcium...	13.00 "
Chloride of magnesia...	4.00 "
Sulphate of lime...	5.00 "
	100.00 "

This spring is said to evolve carburetted or burning gas in considerable quantities, with a small proportion of carbonic acid. It would seem from the composition of its waters to belong to the class of weak brine or salt springs.

Near Catskill, in Greene County, and in Rensselaer County, a mile from the village of Sandlakê, strong *chalybeate springs* are found.

Other springs of the same character are found in Delaware, Dutchess, and Columbia Counties.

CHAPTER XXXVI.

NEW YORK SULPHUR SPRINGS.

Sharon Springs—Avon Springs—Richfield Springs.

WATERS to some extent impregnated with sulphur exist in almost every great section of the State of New York; but few of these springs, however, have been extensively improved for public use, or are so strongly charged with gas and rich in solid medicinal materials as to make them objects of more than local interest. There are, however, several strong exceptions to this general remark, and especially the waters of the Sharon and Avon Springs, which have acquired quite an extended reputation.

As is found to be the case in Virginia, the sulphur springs of New York are generally on, or not very remote from, the lines of fracture or disturbance in the strata of the earth from subterranean causes. The Sharon is said to be the strongest exception to this general law of their location.

Mr. Hall, who made a geological survey of a portion of this State, remarks that springs which issue from different classes of rock are marked by a general character and aspect which indicate their relative geological positions. "In the strata of the Niagara group the water has usually a dark appearance in the spring, though it is limpid and differs essentially from the waters of the salt group, while in higher rocks it is not only less copious, but it is often marked by a black and red deposit, as well as sometimes a whitish stain upon the rock or at the bottom of the spring." These springs, however widely separated, have been observed to have a temperature somewhat above the common springs of their neighborhood. The same fact has been observed in reference to the sulphurous springs so abundantly found in Virginia, going to show a common cause for the general thermalization of such waters.

SHARON SPRINGS.

These springs are in the county of Schoharie, and near the village of Leesville. According to Dr. Beck, they arise from pyritous slates, underlying strata of Helderberg limestone.

The two springs most noted are called *White Sulphur* and *Magnesia.*

The *White Sulphur* has been analyzed by Dr. J. R. Chilton, of the city of New York, who obtained the following results from one pint of the water:—

Sulphate of magnesia....................................	2.65 grains.
Sulphate of lime..	6.98 "
Chloride of sodium.......................................	0.14 "
Chloride of magnesium................................	0.15 "
Hydrosulphuret of sodium } Hydrosulphuret of calcium }	0.14 "
	10.06 "

Sulphuretted hydrogen gas, 1 cubic inch.

Dr. Beck remarks "that sulphate of lime in small fresh perfect crystals is found near the springs in considerable abundance."

Dr. Bell remarks, after quoting the analysis given above, that the "solid contents of a gallon of this water,* as determined by the same chemist, are 160.94 grains, and the amount of sulphuretted hydrogen gas

---

* Mineral and Thermal Springs.

16 inches. The results, as reported by Dr. North, are at variance with the preceding table of reduction to a pint made by Dr. Beck, still from Dr. Chilton's analysis."

The *Magnesia Spring*, according to the analysis of Professor Reed, of New York, contains the following ingredients in one gallon of water:—

Bicarbonate of magnesia	30.5 grains.
Sulphate of magnesia	22.7 "
Sulphate of lime	76.0 "
Hydrosulphates of magnesia and lime	0.5 "
Chloride of sodium and magnesia	3.0 "
	132.7 "

Sulphuretted hydrogen gas, 3.3 inches.

In looking to the relative character of the Sharon waters, we find them most to resemble the *Avon Springs* of New York, and the *White Sulphur Springs* of Virginia, and in a general way they will be found adapted to the same class of diseases for which the latter waters are beneficially used.

The hotel accommodations for visitors at Sharon are represented as extensive and agreeable, with pleasant promenades through well-shaded woodlands contiguous to the spring, and the enjoyment of extensive and interesting views of the surrounding country.

Travelers to Sharon, either from the north, east, or south, should make Albany a point where they take the Binghamton Railroad to Palatine Bridge, and thence by stage-coaches over the mountains to the springs.

AVON SPRINGS.

These springs are situated in the western part of the State, on the eastern branch of the Genesee River, and near the village of Avon. They are about eighteen miles from the city of Rochester, and twenty-four from

Canandaigua. They are connected with Rochester by a daily line of stage-coaches. The Genesee Valley canal-boats also land passengers within nine or ten miles of the springs, whence they are conveyed in coaches to their destination.

The Indians of that region, it is said, knew of and appreciated these springs as "medicine-water" many years ago. The first recorded use of them by the white settlers was in 1792, when they were successfully used for a cutaneous affection. In 1795 we hear of their curing rheumatism of long standing, that had resisted successfully the skill of intelligent physicians. The accommodations at and near the springs are very good, and sufficiently extensive for a large number of visitors. These consist of three hotels near the springs, and two at the village of Avon, from which a connection is kept up with the springs by omnibuses.

There were but two springs known at Avon until the year 1835, and these were designated as the *Upper* and *Lower Springs*. About that time a new one was discovered, which is known as the *New Bath Spring*. This new fountain is said to be thirty feet deep, the water in it rising through a calciferous slate.

An analysis of one pint of the water of this spring yields the following results:—

Carbonate of lime... 3.37	grains.
Sulphate of lime... 0.44	"
Sulphate of magnesia 1.01	"
Sulphate of soda.. 4.84	"
Chloride of sodium.. 0.71	"
11.87	"

Sulphuretted hydrogen, 3.91 cubic inches.
Temperature of the water, 50° Fahr.; specific gravity, 1.00356.

The *Upper*, or, as it is now called, the *Middle Spring*, is about one hundred and fifty yards from the one just described. Its temperature is 51° Fahrenheit.

An analysis of one pint of the water, according to

the investigations of Professor Hadley, shows the following results:—

Carbonate of lime...	1.00 grains.
Sulphate of lime...	10.50 "
Sulphate of magnesia......................................	1.25 "
Sulphate of soda..	2.00 "
Chloride of sodium...	2.30 "
	17.05 "
Sulphuretted hydrogen....................................	12.00 "
Carbonic acid..	5.60 "
	17.60 "

The *New Spring*, Dr. Salisbury states, was formerly a large pool some fifty feet in diameter, and served as a bathing-place for the early inhabitants. It has been more prized as a curative agent than the others, and is more largely resorted to.

In one pint of this water Dr. J. R. Chilton found—

Carbonate of lime...	3.58 grains.
Chloride of calcium..	1.05 "
Sulphate of lime...	7.17 "
Sulphate of magnesia......................................	6.21 "
Sulphate of soda..	1.71 "
	19.72 "

Of gaseous contents:—

Sulphate of hydrogen.......................................	1.32 grains.
Carbonic acid..	0.50 "
Nitrogen..	0.67 "

And a minute fraction of atmospheric air.

This is a uniform and very bold spring, discharging at every season of the year about fifty-four gallons a minute. Its temperature is 45° to 47° Fahr., and its specific gravity 1.0018. Its taste, while decidedly sulphurous, is slightly bitter and saline.

It will be observed that this water contains less sulphuretted hydrogen, and more solid contents, especially of the purging salts, than the Upper or Middle Spring.

In addition to the springs enumerated, there are three

others, called *Iodine* or *Sylvan Springs*, about two miles from the Lower Spring. In these the chloride of sodium strongly predominates, and hence their saltish taste. One of them has but a slight sulphurous impregnation, and somewhat resembles in taste the Congress water after its gas has escaped. We have an analysis of one of these springs, which shows it to contain iodide of sodium, with heavy impregnations of the chlorides of sodium and magnesium, and the sulphate of lime.

In one gallon of the water of this spring Dr. J. R. Chilton found the following ingredients:—

Chloride of magnesium	62.400	grains.
Chloride of sodium	97.440	"
Sulphate of lime	80.426	"
Carbonate of magnesia	15.974	"
Carbonate of lime	26.800	"
Vegetable matter	.240	"
Iodide of sodium.		
	296.240	"
Sulphuretted hydrogen	20.684	cubic in.
Carbonic acid	4.992	"
	25.676	"

In comparing the waters of these springs with the waters of the White Sulphur, in Virginia, it will be observed that the former contains an appreciably larger quantity of lime than the Virginia springs, and that their sulphate of soda and sulphate of magnesia are somewhat in excess of the Virginia waters. The chloride of sodium, existing so largely in the Iodine or Sylvan Spring, and to an appreciable extent in the Lower and New Spring, is discovered only in the very minute portion of about half a grain to the gallon in the Virginia White Sulphur.

Many peculiar operative effects of these waters, as noticed by Dr. Salisbury in his valuable little work on the Avon Springs, are strikingly the same that I noticed in this and the early editions of my work as distinguishing the operations of the White Sulphur waters. Among the most striking of these are the facts noticed by Dr. S. of the similarity of the action of these waters and that of calomel, and that they sometimes produce copious salivation. As is the case with the White Sulphur, the most valuable effects of the Avon waters are found in their *alterative* or *changing effects*, and these effects are best promoted by using them in such doses as do not much increase the natural evacuations of the body. Like the White Sulphur, the quantity of sulphuretted hydrogen gas which the Avon waters contain is too large for its kindly effects in many cases, and hence Dr. S. remarks that after it has been heated, and therefore deprived of a portion of its gas, it becomes more aperient, and that it may be used in this way "when the inflammatory diathesis prevails to such an extent as to resist its beneficial and successful administration in its natural state." The proper *graduation of the amount of sulphuretted hydrogen gas* to the wants and ability of the system to bear it, especially in commencing the use of the water, is a practical matter of great importance in the use of such waters, and one to which I have directed a careful attention for many years.*

The RICHFIELD SPRINGS are in the county of Otsego. They are waters that have come into popular notice within the last few years, and are now largely visited.

The analysis of these waters by Prof. Reed shows that one gallon of the water contains—

Bicarbonate of magnesia	20 grains.
Bicarbonate of lime	10 "
Chloride of sodium and magnesia	1.5 "
Sulphate of magnesia	30 "
Hydrosulphate of magnesia and lime	2 "
Sulphate of lime	20 "
Sulphuretted hydrogen gas per gallon	26.6 inches.

---

* See chapter vi., on the Relative Virtues of the Saline and Gaseous Contents of the White Sulphur Water, etc.

CHAPTER XXXVII.

NEW YORK SULPHUR AND ACIDULOUS SPRINGS.

Clifton Springs—Chittenango Springs—Messina Sulphur Springs—Manlius Springs—Auburn Springs—Chappaqua Springs—Harrowgate Spring—Spring at Troy—Newburg Spring—Springs in Dutchess and Columbia Counties—Catskill Spring—Nanticoke Spring—Dryden Springs—Rochester Spring—Springs in Monroe County: Gates, Mendon, and Ogden—Verona Spring—Saquoit Springs—Springs in Niagara County—Seneca or Deer Lick Springs—Oak Orchard Acid Springs—Byron Acid Springs—Lebanon Spring—Adirondack Spring.

IN addition to the two principal sulphurous springs of Sharon and Avon already noticed, there are numerous others of less public notoriety. The first of these I shall mention is—

CLIFTON SPRINGS.—They are situated in the county of Ontario, between Vienna and Canandaigua. In importance they should, probably, rank next to Sharon and Avon. The odor and taste of these waters are distinctly sulphurous. Their temperature is 51° Fahr. These waters, Dr. Beck asserts, have their origin in hydraulic limestone, underlying a stratum of common limestone. There are here several springs, one of which is very bold and yields a large amount of water. No analysis of these waters has been given to the public, that I am aware of.

CHITTENANGO SPRINGS are in the county of Madison, near Chittenango Creek. Two springs here have attracted attention; their temperature is 49° Fahr. They have been ascertained to contain the sulphates and carbonates of lime, sulphate of magnesia, chloride of sodium, with sulphuretted hydrogen and carbonic acid gases. Dr. Beck remarks that these waters are highly esteemed in many cases of disease, and, their location being very eligible, he expresses the opinion that when they are better known they will be much resorted to.

MESSINA SULPHUR SPRINGS are situated three miles northeast of Syracuse, and one mile from the Erie Canal. The temperature of their water is 50° Fahr., and its taste strongly sulphurous. It is said to have been used with good effects in many cases.

An analysis of the water shows it to contain, in one pint—

Carbonate of lime	1.85 grains.
Sulphate of lime	8.55 "
Sulphate of magnesia	1.36 "
Chloride of calcium	1.33 "
	13.09 "

MANLIUS SPRINGS are situated in Onondaga County. They are slightly saline in taste, and are impregnated in but slight degree with sulphuretted hydrogen gas. They have acquired some local reputation as a remedial agent.

In the neighborhood of these springs there is a small sulphurous lake, known by the name of *Lake Sodom*. We are told by Dr. Beck that the depth of this lake gradually increases from its northern outlet from twenty-five to one hundred and sixty-eight feet, and that water drawn from this depth is found to be highly impregnated with sulphuretted hydrogen. The color of the water in this lake is a deep green, from which it is sometimes called *Green Pond*.

AUBURN SPRINGS.—There are two springs that bear this name, separated several miles from each other.

One of these is situated two miles north of the village of Auburn; the other four miles west of the same village. An analysis of the latter spring, by Dr. Chilton, shows the following ingredients in one pint of the water:—

Sulphate of lime	15.00 grains.
Sulphate of magnesia	3.20 "
Chloride of magnesium	0.25 "
Chloride of, sodium	0.75 "
	19.20 "

Sulphuretted hydrogen, 1.5 cubic inches.

In the valley of the Hudson, Dr. Beck mentions numerous sulphurous springs. They are found from the neighborhood of Sing Sing to Fort Miller, a distance of one hundred and fifty miles.

The CHAPPAQUA SPRING is four miles from Sing Sing. It holds in solution sulphate of lime, chloride of calcium, and the muriates of iron and magnesia.

HARROWGATE SPRING is near Greenbush, in Rensselaer County.

There is also a sulphurous spring in the northern end of the city of Troy, in Rensselaer County.

There are several sulphur springs in the county of Albany, one of them very near the city of Albany.

The NEWBURG SPRING, slightly impregnated with sulphuretted hydrogen, is in the county of Orange.

In Dutchess and Columbia Counties there are several springs. The most noted one in Dutchess is near Ameniaville. In Columbia there is one on the farm of Mr. McNaughton, between* the Shaker Village and the Lebanon Springs, and another near the village of Kinderhook.

The CATSKILL SPRING is two miles from the village of Catskill, in the county of Greene. There are several others in the same neighborhood.

In the southwestern part of the State we find the *Nanticoke Spring*, in the county of Broome. It has acquired considerable reputation. *Dryden Springs* are in the town of Dryden, in Tompkins County, ten miles from Ithaca. They have acquired reputation in their region of country, and are considerably resorted to.

ROCHESTER SPRING, otherwise known as *Longmuir's Sulphur Well*, in the city of Rochester, is much used by the inhabitants of the city. It rises through a boring of two hundred feet in depth. It deposits, when heated to 100° Fahr., carbonate of lime and sulphur. Its temperature at the surface is usually 52° Fahr., and its specific gravity 1.00407. One pint of the water contains—

Carbonates of lime and magnesia, with a trace of iron.	1.48 grains.
Chloride of sodium	6.52 "
Sulphate of soda	6.99 "
	14.99 "

Sulphuretted hydrogen, 2.16 cubic inches.
Carbonic acid in small quantity.

In the county of Monroe are the sulphurous springs of *Gates*, *Mendon*, and *Ogden*, at all of which there are suitable bathing arrangements and proper accommodations for visitors.

Verona Spring is in Oneida County, fourteen miles from Utica. Professor Noyes's analysis of the water of this spring shows that one pint contains—

Chloride of calcium, with chloride of magnesium...	8.50 grains.
Sulphate of lime	7.50 "
Chloride of sodium	90.00 "
	106.00 "

Sulphuretted hydrogen is very abundant in the water, amounting almost to complete saturation.

About nine miles from Utica are the *Saquoit Springs*. Their waters are very highly impregnated with carburetted hydrogen, and contain, in considerable quantities, the chlorides of sodium and magnesium, with a small portion of the sulphate of lime and a trace of iron. So abundant is the carburetted hydrogen in the water, that it is collected, conducted through tubes, and kept constantly burning.

In Niagara County there are several sulphur springs; among them may be mentioned those near the Falls of Niagara, those near Lockport, and those also in the neighborhood of Lewistown.

The *Seneca* or *Deer Lick Springs* are in Erie County, four miles from Buffalo. They hold in solution carbonates of lime, soda, and magnesia, with sulphate of lime. They abound richly in sulphuretted hydrogen.

We are told that sulphurous springs are also found in the northern part of New York, in Lewis, Clinton, and St. Lawrence Counties.

ACID SPRINGS OF NEW YORK.

In addition to the acidulo-saline and sulphurous waters already described, there are in New York several *acidulous* springs. The acid quality of these waters is owing to their holding in solution an excess of sulphuric acid, which is readily detected both by their taste and by chemical reagents.

ACID SPRINGS OF NEW YORK.

These waters are found to be so largely impregnated with iron in the form of a protosulphate, and with sulphate of *alumina*, as to entitle them to be called *chalybeate* or *alum* waters with as much propriety as they are called *acidulous*. Similar springs in Virginia are uniformly known by the name of *alum springs*.

Acidulated aluminous springs, partaking of the same general character of the acid springs of New York, which we are about to consider, are found in every neighborhood in certain geological districts in Virginia, and especially on the eastern and western slopes of the Alleghany chain of mountains, through the entire district there known as the great "Spring Region."

Fountains of the same general character are found in Pennsylvania, and also in the eastern portion of Tennessee, and will probably be discovered along the entire course of the great Appalachian upheavings, or axis of disturbance from the extreme north to the alluvial plains of the Gulf of Mexico.

The principal springs of this class in New York are the *Oak Orchard Springs*. They are eight in number. Their situation is in Genesee County, eight miles southeast from Lockport, and about six miles from the Erie Canal, at the village of Medina. These waters have been analyzed by Professor Emmons and Dr. Chilton.

Professor Emmons's examination of Spring No. 1 shows that one pint of the water contains the following ingredients:—

Free sulphuric acid	31.50 grains.
Sulphate of protoxide of lime	19.50 "
Sulphate of lime	4.50 "
Sulphate of magnesia	2.00 "
Silica	0.33 "
Organic matter	1.33 "
	59.16 "

Equal to 473.28 grains to the gallon.

Spring No. 2 was found to contain but 24.25 grains of free acid and solid ingredients to the pint, and No. 3 but 19.33.

Dr. Chilton, by an analysis of one gallon of the water of Spring No. 1, arrives at results strikingly different from those of Professor Emmons. His researches* show one gallon to contain—

Free sulphuric acid	82.96 grains.
Sulphate of lime	39.60 "
Phosphate of iron	14.32 "
Sulphate of alumina	9.68 "
Sulphate of magnesia	8.28 "
Silica	1.04 "
Organic extractive matter	3.28 "
	159.16 "

Equal to about 10 grains to the pint.

The difference in the amount of these waters in the several fountains during wet and dry weather is always noticeable, and in some instances is very remarkable. Generally they are surface springs, the waters obtaining their peculiar impregnations by percolating through the peculiar argillite slate in which they are found. Whatever difficulties there may be in accounting for the peculiar impregnations of some mineral waters, there are none in reference to this class, for portions of the slaty rock through which the waters percolate, when immersed in common water, produce the very same impregnations that are found in the water in the pools in which it is collected for use. Many persons in the South use at their homes the Virginia alum waters prepared in this *pro re nata* way from the rock obtained from the various alum fountains.

Taking Dr. Chilton's analysis as the standard, the *Oak Orchard Springs* more resemble the Rockbridge alum waters in Virginia than any others to which they

---

* Mineral and Thermal Springs of the United States and Canada.

can be compared. The resemblance is only striking in this, however, that they both contain free sulphuric acid, alumina, and iron in marked proportions; the sulphuric acid, lime, iron, and magnesia in the New York springs being greater than in the Virginia waters, while the alumina and silica are more than fifty per cent. greater in the latter. In addition to these ingredients, common to both waters, the Rockbridge Spring contains chlorate of sodium, crenate of ammonia, and free carbonic acid, ingredients not found in the Oak Orchard Springs.

As therapeutic agents, this class of waters are tonic and astringent. In enfeebled conditions of the digestive and uterine functions,—in cases of pure *atony* or *feebleness* unaccompanied by inflammation or irritation in any of the organs,—in exhaustion from previous disease, where the chief complaint is debility,—and in cases of *anæmia* or poverty of the blood, when unconnected with obstinate visceral obstructions, they are safely and beneficially prescribed. In passive hemorrhages, long-continued intermittents, and dropsical effusions, unattended with organic obstructions, and in leucorrhœa and chlorosis, they are often beneficial. In chronic diarrhœa, as well as in chronic irritations and debility of the kidneys, bladder, and urethra, they are usefully employed. The Virginia waters of this class have proven eminently remedial in scrofula; indeed, no remedy is now attracting so much attention for this formidable disease, in the Southern country, as the alum waters. Upon this particular subject, as well as for a more general notice of the therapeutic range of such waters, I refer the reader to what has been said under the head of the *Rockbridge Alum Springs*.

Dr. S. P. White[*] thinks favorably of the Oak Orchard Spring waters in some of the cutaneous dis-

---

[*] Paper read before the New York Academy of Medicine in December, 1848. *Vide* " Mineral and Thermal Springs of the United States," etc.

eases, and in the colliquative sweats of hectic fever. He regards it as worthy of a trial in the phosphatic diathesis, in colica pictonum, and asthma, and also in chronic laryngitis, pharyngitis, and chronic conjunctivitis.

Dr. White recommends that this water be taken in "about a wineglassful, diluted with simple water, three times a day." This dose is much smaller than I have been accustomed to recommend in the use of similar waters. The practice found most beneficial with the Virginia waters of the same general character is to use from two to six half-pint glasses in the course of the twenty-four hours.

At Clifton Springs, twelve miles from Geneva, there is an acid spring. I have not seen an analysis of it.

BYRON ACID OR SOUR SPRINGS are the names given to two acidulous springs in the town of Byron, Genesee County. One of these springs is near the Byron Hotel, and is remarkable for the great quantity of acid contained in its waters. It is a stream of considerable boldness, so much so as to be sufficient to operate a grist-mill.

Dr. Beck describes this water as intensely sour, transparent and colorless, and of the specific gravity of 1.11304 at 60° Fahr. Its saline matter, which is small, consists of silica and alumina, with a small quantity of oxide of iron and lime. Dr. Beck remarks that "this is a nearly pure, though dilute, sulphuric acid, and not a solution of acid salts as has been supposed, for the bases are in too minute a proportion to warrant the latter opinion."

LEBANON SPRING belongs to the thermal class of waters. It is in the county of Columbia. The bathing here is very delightful, the temperature of the water being constantly 73° Fahrenheit. Its mineral impregnation is scarcely noticeable, being only a grain and a quarter in a pint. So abundant is the supply of this thermal water that it is employed to operate two or three mills erected at no great distance from its source.

The ADIRONDACK SPRING was discovered in 1868. It is situated in the village of Whitehall, forty miles north of Saratoga, at the head of Lake Champlain.

It has been analyzed by Prof. C. Collier, of Vermont University, who reports that one gallon of it contains—

Acids.

Carbonic acids, free	31.861 grains.
Carbonic acids, combined	22.591 "
Sulphuric acid	6.594 "
Chlorine	8.701 "

Bases.

Oxide of iron	3.129 "
Oxide of manganese	traces.
Lime	14.970 "
Magnesia	7.914 "
Alumina	traces. "

Alkalies.

Potassa	3.623 "
Soda	10.602 "
Lithia	.009 "
Silica	.742 "
	110.696 "

Temperature, 52° Fahr.

Sulphate of lime	11.134 "
Carbonate of lime	18.543 "
Carbonate of magnesia	16.618 "
Carbonate of iron	5.040 "
Carbonate of manganese	traces.
Carbonate of potash	5.317 "
Carbonate of soda	5.135 "
Carbonate of lithia	.023 "
Chloride of sodium	14.340 "
Alumina	traces.
Silica	.742 "
Free carbonic acid	67.275 cubic in.

The analysis shows the water to be a *saline chalybeate*, and of promising therapeutic character.

CHAPTER XXXVIII.

SPRINGS OF PENNSYLVANIA.

Bedford Springs—Gettysburg Spring—Frankfort Mineral Springs—Chalybeate Spring near Pittsburg—York Springs—Carlisle Springs—Perry County Springs—Doubling Gap and Chalybeate Springs—Fayette Spring—Bath Chalybeate Spring—Blossburg Spring—Ephrata Springs—Yellow Springs—Caledonia Springs.

PURSUING the plan I have adopted of introducing the States somewhat in respect to the extent and importance of their mineral waters, I next notice the mineral springs of Pennsylvania; and first, as holding the highest rank among her mineral fountains, the

BEDFORD SPRINGS.

The strong mineral impregnation of the Bedford waters, their valuable therapeutic effects, the high mountain altitude in which they are situated, together with the delightful summer climate and very pleasant mountain scenery of their neighborhood, combine to make them a place of large, pleasant, and useful resort, alike to the seekers of health and the votaries of pleasure. They are in the county of Bedford, and two miles from the village of Bedford, one hundred miles west of Harrisburg, and one hundred and thirty miles northwest from Baltimore; they are less than one hundred miles east of Pittsburg, and one hundred and thirty northwest from Washington.

The principal spring is known as *Anderson's;* the

others are called *Sweet, Sulphur, Chalybeate, Limestone,* and *Fletcher's* or *Upper Spring.*

ANDERSON'S SPRING is a saline chalybeate water. Its most active ingredients are sulphate of magnesia and carbonate of iron; the former exists in the water in the large proportion of 80 grains to the gallon, the latter in that of 5 grains. Dr. Church, who analyzed this water in 1825, states that "the water is clear, lively, and sparkling. At 10 A.M. on the 28th of May, the temperature of the water in the spring was 58° Fahr., while that of the surrounding atmosphere was 73° of the same scale. Its specific gravity is 1029. It has a peculiar saline taste, resembling a weak solution of Epsom salts in water, impregnated with carbonic acid, and exhales no perceptible odor. On exposure in an open vessel to the air, it becomes vapid, but does not become turbid or deposit a sediment. The water deposits carbonate of iron on those substances over which it constantly flows. Limestone, iron ore, calcareous and silicious substances abound about the spring."

Dr. Church's analysis of one quart of the water shows the following results:—

Sulphate of magnesia, or Epsom salts...................	20 grains.
Sulphate of lime...	3¾ "
Muriate of soda...	2½ "
Muriate of lime...	¾ "
Carbonate of iron...	1¼ "
Carbonate of lime..	2 "
Loss...	¾ "
	31 "

Carbonic acid gas, 18½ cubic inches.

The SWEET SPRINGS, according to Dr. Church, "are two in number, and issue from fissures in slate rocks, about fifty yards apart, on the east side of Federal Hill, about one hundred and fifty yards from Anderson's Spring, from which they are separated by Shover's Creek. They are copious springs, of remarkably pure water, which is very clear and colorless. Its temperature was, on the 28th of May, 52° Fahr. The water of these springs is used for cooking, washing, etc. by the residents at Bedford Springs, and the visitors decidedly prefer it for drinking-water, and, on account of its purity, they very appropriately called these springs the *Sweet Springs.*"

The SULPHUR SPRING is on the west side of Shover's Creek, about two hundred yards from Anderson's Spring. It is not so copious in its flow as the other springs. Its temperature is 56° Fahr., and it has a strong odor of sulphuretted hydrogen. Dr. Church's experiments with this water determined that it holds in solution carbonic acid, sulphuretted hydrogen gas, with lime, magnesia, and common salt in small quantities. This spring contains no iron.

The CHALYBEATE SPRING, Dr. Church states, "rises in a meadow, about one and a half miles northeast of Bedford, and about three miles from Anderson's Spring. It is not a copious spring. The water exhales the peculiar odor of sulphuretted hydrogen gas, and is covered with a thin whitish pellicle. When first taken from the spring it is clear and limpid, but on exposure in an open vessel to the action of the air it becomes turbid. Its taste is ferruginous and slightly hepatic." Experiments prove that this water contains sulphuretted hydrogen and carbonic acid gases, carbonate of iron, muriate of soda, and a minute portion of magnesia. In cleaning out this spring, many years ago, a part of the skeleton of a mammoth was found imbedded in the mud.

The LIMESTONE SPRING is a bold fountain of pure water, about two hundred yards below Anderson's Spring. Its temperature is 51° Fahr.

FLETCHER'S, or the UPPER SPRING, is on the west side of Constitution Hill, one hundred and fifty yards from Anderson's Spring. Its temperature is 55° Fahr. Dr. Church's experiments with this water show that, as compared with that of Anderson's Spring, it contains rather more iron and common salt, with less magnesia, and about the same proportion of the other ingredients.

The Bedford waters are laxative and tonic in their effects. They are said to "give rise to full purging, and cause a discharge of bilious or other acrid matters, with as much activity as the most powerful purgatives. They also excite the action of the kidneys and skin, causing a very free secretion of urine and perspiration."

GETTYSBURG SPRING.

This spring is located one mile west of Gettysburg, Adams County, Pennsylvania. It rises in a portion of the battle-ground made famous by one of the most sanguinary struggles of our recent civil war. If not entirely unknown as a water possessing curative powers, it was unknown to the fame it now enjoys, until after its surrounding hills were moistened by the fraternal blood of contending hosts.

The legend of the place, that the virtues of the waters were first demonstrated by wounded soldiers who fell in battle in the vicinity of the springs, is probably more romantic than true. But there is little doubt that the distinction which the famous battle gave to the ground in which the spring is situated did much to direct attention to the spring and to lead to a more thorough examination of its waters. The chemical examination of this spring by Prof. Genth recognizes the following constituents, in the proportions stated, in 331 cubic inches of the water:—

THE GETTYSBURG SPRINGS HOTEL.

Sulphate of baryta	trace.
Sulphate of strontia	0,004.27 grains.
Sulphate of lime	0,831.45 "
Sulphate of magnesia	6,779.40 "
Sulphate of potash	0,208.36 "
Sulphate of soda	2,467.76 "
Chloride of sodium	0,657.90 "
Chloride of lithium	trace.
Bicarbonate of soda	0,704.57 "
Bicarbonate of lime	16,408.15 "
Bicarbonate of magnesia	0,542.60 "
Bicarbonate of iron	0,035.85 "
Bicarbonate of manganese	0,006.69 "
Bicarbonate of nickel	trace.
Bicarbonate of cobalt	trace.
Bicarbonate of copper	0,000.50 "
Borate of magnesia	0,034.92 "
Phosphate of lime	0,006.79 "
Fluoride of calcium	0,009.54 "
Alumina	0,003.80 "
Silicic acid	2,030.88 "
Organic matter, with traces of nitric acid, etc...	0,708.70 "
Impurities suspended in the water, like clay, etc.	1,100.69 "
	32,542.72 "

Analyses of mineral waters, however perfect they may be (and they are very often imperfect), cannot, in the very nature of the case, do more than plausibly indicate probable applicabilities and efficiencies, and should always be regarded rather as plausible *hints* to the invalid and the medical man than as positively determining medicinal efficacy and value. The very *tests* which reveal some qualities in mineral waters may have the power of destroying others, and these, too, may be those in which the medicinal properties reside. The remedial properties, then, of mineral waters *cannot be determined with any positive certainty by analysis*, however nicely conducted, but must be ascertained by experience.

One dozen carefully diagnosed and "well-watched" cases, under the use of a mineral water, will do more to determine its medicinal powers than any analysis that can be made by the ablest chemist. But taking the indications which analysis reasonably supposes, and applying them to the Gettysburg water, we would expect to witness from its use the same character of effects that have been known for centuries to result from the European waters of similar chemical composition, such, for instance, as the well-known waters of *Ems*, *Teplitz*, *Mont d' Or*, and *Vichy*, which the Gettysburg water sufficiently resembles to justify a plausible inference that its medicinal efficacy would be similar.

This water belongs to the *carburetted class* of waters, and holds in solution ingredients that have given much reputation to such waters, both in Europe and America; and if judged alone by its analysis, favorably impresses the medical mind as to its therapeutic efficiency in some important forms of disease.

The *bicarbonate of lithia*, found in this spring, is an interesting fact. In addition to the modifying influences which this agent may, and probably does, exert upon its associated ingredients, its affinity for *uric acid*, and its consequent specific efficacy in dissolving uritic concretions when removed from the body, plausibly indicate its adaptation to the same end when internally used.

The chemical composition of this water as shown by its analysis, taken as a whole, plainly indicates its adaptation as a valuable remedy in a long list of affections of the *mucous surfaces*, and especially in *dyspeptic depravities*, *chronic irritations* of the *bowels*, as well as of similar conditions of the *kidneys* and *bladder*. The invalid may hopefully look to the use of waters containing the salts found in the Gettysburg spring, not only in chronic inflammations and irritations of the organs alluded to, but also in certain *pulmonary disturbances*, as *bronchitis*, *chronic laryngitis*, *humoral asthma*, *catarrh*, etc.

The reported curative effects of the water during the last five years are highly favorable to its employment in *dyspepsia*, *chronic diarrhœa*, *gout*, *chronic rheumatism;* in the various *kidney and bladder affections*, and especially in those of *uric acid* predominance. In *albuminuria* or *Bright's disease* it has been successfully prescribed before positive degeneration of the kidneys had taken place, and in some cases of *diabetes* it has been successfully prescribed.

This water is decidedly *alterative*, as well as *specific;* indeed, its principal sanative influences are exerted in its *alterative power*. This supposes its absorption into the general current of the circulation, and the influence there of the efficient medicinal materials which it holds in solution, in correcting the blood and the diseased organs and tissues, which such medicinal materials are adapted to alterate and correct; thus bringing them into a natural performance of their functions, and imparting a healthful tone and energy to the whole system.

The operations of this and all other *alterative waters* are quiet and unobtrusive; *slow*, but all the more *sure, and permanently valuable, because slow, in radically and effectually accomplishing their important mission.* Its immediate and sensible effects are not very marked, producing ordinarily but little effect upon any of the excretory organs.

This water, although very extensively employed in the treatment of disease for the last five years, has been mostly used as *transported water*, and in many instances after it had been removed from the spring several months. Recently, extensive improvements have been made at the spring, capable of accommodating a large number of visitors.

The Gettysburg water, being essentially *ungaseous*, and *holding its salts firmly in solution*, is exceedingly well adapted for *transportation*. Indeed, with the single exception of its parting with that *earthy freshness* peculiar to all waters just issuing from their source, *it undergoes no change by transportation*, either by deposition of its salts, taste, general appearance, or medicinal efficacy. This is a valuable feature in the water, and while it increases the material value of the fountain to its proprietors, gives confident assurance to invalids of the equal efficacy of the transported water with that used fresh at the spring.

A proper method of using the Gettysburg water in ordinary cases is to take from one and a half to three pints in the course of the day and night,—that is, from one to three half-pint glasses at intervals before breakfast, one before dinner, and from one to two before retiring at night.

FRANKFORT MINERAL SPRINGS.—These springs are situated in Beaver County, twenty-six miles southwest from Pittsburg, and one mile and a half from the village of Frankfort. The principal spring is known as *Cave* Spring. It arises within a large and very romantic cave, on the plantation of Mr. Stevens. The cave itself is an interesting natural curiosity, and is much visited by the people of the surrounding country. Dr. Church, of Pittsburg, directed attention to the medicinal virtues of the Cave Spring water many years ago. By his analysis the water is found to contain carbonic acid, carbonate of iron, carbonate of magnesia, muriate of soda, a minute portion of bitumen, and sulphuretted hydrogen gas.

There is a fountain known as *Leiper's Spring*, very near Frankfort, which Dr. Church found to hold in solution somewhat more carbonate of iron and muriate of soda, with less magnesia, and about the same proportion of carbonic acid, sulphuretted hydrogen, and bitumen, that is found in the Cave Spring water.

Dr. Church remarks that these waters sometimes occasion nausea and vomiting when first drunk, but generally they set kindly and pleasantly on the stomach. It generally operates mildly on the bowels and copiously by the kidneys. With some persons its free use occasions vertigo, with slight sensation of intoxication. As a therapeutic agent, it is said to " regulate the bowels, strengthen the stomach, improve the appetite, clear the skin, promote diaphoresis, and cause great freedom of urination."

CHALYBEATE SPRING NEAR PITTSBURG.—This spring is about four miles from the city of Pittsburg. Dr. John Bell* gives the following description and analysis of it by Dr. Meade:—

"When the water remains undisturbed for a few hours, it is covered by a white pellicle, its taste is lively and rather pungent, with a peculiar ferruginous flavor, and it exhales an odor of sulphuretted hydrogen gas. Its temperature is very generally uniform, and is 54° Fahr. The specific gravity of the water differs little from the purest water, and is as 1.002 to 1.000.

"According to Dr. Meade's analysis, it contains muriate of soda, 2 grains; muriate of magnesia, $\frac{1}{4}$ grain; oxide of iron, 1 grain; sulphate of lime, $\frac{1}{2}$ grain; carbonic acid gas in one quart of water, 18 cubic inches.

"Dr. Meade thinks this water even superior, in a medical point of view, to the water of the *Schooley's Mountain Spring*, which has long sustained a high character for its chalybeate properties."

YORK SPRINGS.—These springs are in Adams County, and are readily reached by railroad from Philadelphia and Baltimore. There are here two principal springs, one strongly *chalybeate*, the other distinctly *saline*. The latter contains 6 grains sulphate of lime, 4 muriate of soda, and 1.20 sulphate of magnesia in a pint of water. This spring is said to be diuretic and somewhat cathartic. The chalybeate is doubtless adapted to the class of diseases in which chalybeate waters are commonly prescribed.

---

* Mineral and Thermal Springs, etc.

Carlisle Springs are mild *sulphurous waters*. They are near the pleasant town of Carlisle, through which passes the railroad from Philadelphia to Pittsburg. The hotel accommodations here are said to be very good.

Perry County Springs.—These springs are at the base of Pisgah Mountain, fourteen miles from Harrisburg, and eleven from Carlisle. They belong distinctly to the *thermal class*, their temperature being from 70° to 72° Fahr. When used as a drink they are gently aperient and decidedly diuretic. They are most esteemed as a bath, and employed in this way have proved beneficial in various disorders, and especially in diseases of the skin.

Doubling Gap Sulphurous and Chalybeate Springs.—These springs are in Cumberland County, about thirty miles west from Harrisburg. They are eight miles from Newville, through which the Cumberland Valley Railroad passes, and from whence passengers to the springs are conveyed by stages.

I am indebted to Dr. John Bell for Professor Booth's chemical examinations of these waters. He says, "The odor of sulphuretted hydrogen, perceived at some distance from the springs, imparts to this water the peculiar properties of sulphur springs. Besides this ingredient, I find that the water contains carbonates of soda and of magnesia, Glauber's salts, Epsom salts, and common salt; ingredients which give it an increased value. After removing the excess of carbonic acid which it contains, it gives an alkaline reaction."

Of the other springs he remarks, "The chalybeate water readily yields a precipitate after ebullition or continued exposure to the excess of carbonic acid. Besides the bicarbonate of iron, which is the chief characteristic, it also contains Epsom salts, common salt, and carbonate of magnesia."

The composition of these springs indicates with sufficient clearness their respective applicability as therapeutic agents. The first belongs to the mild sulphurous saline, the second to the carbonated ferruginous class.

Fayette Spring.—This spring is situated on the eastern slope of the Laurel Hill, and near the great National road. The water is chalybeate, very cold and abundant in quality. The scenery around the spring is wild and romantic, and the coolness, freshness, and elasticity of the air wholesome and invigorating.

Bath Chalybeate Spring is near the town of Bristol, on the Delaware. Dr. Bell informs us that "these springs used to be visited by many of the citizens of Philadelphia, on account, in good part, of ready access to them," and that Dr. Benjamin Rush wrote an account of them in 1773. They seem now to have gone very much out of public notice.

BLOSSBURG SPRINGS.—These springs belong to the class known as acid waters in New York, and as alum springs in Virginia. In taste they very much resemble the Rockbridge Alum water. They contain a large amount of free sulphuric acid, and less alumina than the Virginia waters. Unlike Rockbridge water, they readily deposit, when removed from the spring, a large portion of the iron they hold in solution.

The Blossburg waters are adapted to the same general class of diseases for which the Virginia and New York acid waters are beneficially prescribed. The dose of a "tablespoonful," in which they are sometimes recommended, is altogether too small to produce any beneficial effects in ordinary cases. I have had an opportunity of examining the Blossburg waters, and of carefully comparing them with the Rockbridge waters, and I am sure, judging from the relative strength of the two, and from my knowledge of the proper dose of the latter, that from two to four or even five glasses of the Blossburg waters may in many cases be beneficially taken in the course of the twenty-four hours.

These springs are in Tioga County, near the New York line, and in the immediate region of beds of iron and bituminous coal.

In addition to the mineral springs of Pennsylvania, already noticed, there are numerous pure, cool, and invigorating fountains, that from the great purity of their waters, their healthful situation, the character of their accommodations, and the facility with which they may be reached, have become places of considerable summer resort. In this category may be reckoned the *Ephrata*, *Yellow*, and *Caledonia Springs*. I will notice them in the order in which I have named them.

The *Ephrata Springs*, the annual resort of many persons during the summer season, are situated in the rich agricultural county of Lancaster. The grounds around them are very pleasant, the scenery interesting, and the hotel accommodations excellent. Baths of various temperatures are furnished, and many inducements offered to make the sojourn of visitors at these springs both agreeable and beneficial.

The *Yellow Springs* are thirty miles from Philadelphia, in the county of Chester. From these springs a magnificent view of a most interesting surrounding country is obtained. The rides and drives are very pleasant, and the twice daily communication with Philadelphia by the Reading Railroad and stages offers great facilities to the citizens of the city in the enjoyment of country air and spring recreations. They have facilities here for the shower and douche, as well as for the common immersion baths. The hotel accommodations are said to be most excellent.

CALEDONIA SPRINGS were formerly known as *Sweney's Cold Springs*. They are about fifteen miles from Chambersburg. Visitors to them, on arriving at Chambersburg, may immediately proceed by coach to their destination. The water of these springs, used as a bath, has enjoyed a high local reputation for many years, in the cure of various diseases for which cold, tepid, or warm baths are commonly employed. Chronic rheumatism has been often submitted to the Caledonia bath, and, it is said, with excellent effect.

The waters of Caledonia are very pure, the baths comfortable, the *cuisine* admirable, while the mountain and intervale scenery, and the elastic, invigorating atmosphere, afford all that could be desired of scenery or climate to delight the mind, invigorate the system, and give new life and energy to the *habitués* of cities, worn down in the treadmill of incessant toil, counting-room confinement, or commercial anxieties.

CHAPTER XXXIX.

MINERAL SPRINGS OF VERMONT.

CLARENDON GASEOUS SPRINGS.—This is a mild acidulous water, very slightly impregnated with saline matter, so slightly, indeed, as to make it rank among the purest waters known. Dr. Bell* states on the authority of Dr. Gallup, who published a notice of this spring, that it has been ascertained by analysis to contain in an American gallon, 235 cubic inches, the following ingredients:—

Nitrogen or azote	9.63 cubic inches.
Carbonic acid	46.16 " "
Besides atmospheric air.	
Carbonate of lime	3.02 grains.
Muriate of lime, sulphate of lime, and sulphate of magnesia	2.74 "
	5.76 "

Temperature of the Higher spring 48° Fahr., of the Lower 54° Fahr.

These waters have acquired considerable reputation in the surrounding country for the cure of dropsical effusions, diseases of the skin, chronic bronchitis, irritations of the bladder, etc.

The quantity of the water advised to be used varies from five to twenty-five half-pint tumblers in the course of the twenty-four hours. It is said that on commencing their use they often excite slight nausea, with a sense of warmth on the surface, but that those

---

* Mineral and Thermal Waters, etc.

sensations disappear in five or six hours, in which time their diuretic effects will be manifest.

NEWBURG SULPHUR SPRING is twenty-seven miles in an easterly direction from Montpelier. This is a spring of some notoriety in the country around, and considerably resorted to by invalids. No analysis, so far as I know, has been made of the water, but it is said to be very strongly impregnated with sulphuretted hydrogen gas. Other springs of similar character are found in the same region of country.

There are good hotel accommodations here, and pleasant facilities for bathing. The use of the water has been much praised in diseases of the skin, and in scrofulous affections.

HIGHGATE SPRINGS, eleven miles from the boat-landing at Albon's Bay, are sulphurous waters,.and of the same general character as those of the Newburg Spring.

The ABBURGH SPRING is a sulphurous water, similar to the waters of Newburg and Highgate just noticed.

Professor Hitchcock mentions a thermal spring near Bennington, but does not give its temperature. It throws off oxygen and nitrogen gases, and the water is so abundant that it is used for operating machinery.

MISSISQUOI SPRINGS—VERMONT SPRINGS.

In the neighborhood of the village of *Sheldon* are two mineral springs that have recently been brought into public notice, mainly through the transportation of their waters, and publications claiming for them extraordinary virtues. One of these is known as the *Missisquoi*, the other as the *Vermont Spring*. Rising in the same neighborhood and in the same geological range, and the qualitative analysis of the two being very similar, I assume that the two springs do not essentially differ in therapeutic qualities. They are both shown to contain sodium, calcium, magnesium, manganese, iron, alumina, chlorine, and silica, with sulphuric and hydrochloric acid.

So far as the analyses of these waters indicate their therapeutic powers, their best effects may be looked for in *cuticular diseases, ulcerations, strumous conditions* of the system, and in the *tertiary form of syphilis*. The claim urged in behalf of these waters as a specific for the cure of *scirrhus* or *cancer* requires for its establishment more satisfactory evidence, I conceive, than has yet been given to the public.

Pathology in reference to *cancer* is so often at fault, or, in other words, obstinate disorders far less intractable are so often mistaken for it, that reports of the cure of such cases ought to be received with caution; not because of any intention to deceive on the part of the relater, but because of his liability to be deceived as to the true pathology of such cases. I attach the more importance to this caution, because mineral waters in my hands, however efficacious they have been in skin diseases and in ill-conditioned ulcers, have never been found to be remedial in true scirrhus or cancer.

Medical men, and every friend of humanity, will rejoice in an admitted specific that can be relied upon to cure and eradicate this terrible affection; nor should we hold such results to be impossible, for it is not unreasonable to suppose that nature is capable of providing, perhaps has provided, a physical remedy for all her physical ills.

The ALBURGH SPRING is near Missisquoi Bay, Grand Isle County, sixteen miles from St. Albans. Prof. Chandler, from his chemical analysis, states that this water contains—

Potassium, sodium, lithium, lime, magnesia, strontia, chlorine, sulphuric acid, carbonic acid, and silica.

These exist in the form of the following compounds:—

Chloride of potassium, chloride of sodium, sulphate of potassa, bicarbonate of lithia, bicarbonate of soda, bicarbonate. of lime, bicarbonate of strontia, bicarbonate of magnesia, and silica.

The most interesting feature of this water, as exhibited by its analysis, is its distinct alkaline character, and the presence in it of the carbonate of lithia.

CHAPTER XL.

SPRINGS OF MASSACHUSETTS.

HOPKINTON SPRINGS have acquired some reputation in the section of country in which they are situated. An analysis of the water of the principal spring, by Dr. Gorham, shows that it contains the carbonates of magnesia, lime, and iron. One of the springs here is strongly impregnated with sulphur.

BERKSHIRE SODA SPRING.—This watering-place is situated in the mountains in Berkshire County, three miles from the village of Great Barrington, through which the cars of the Housatonic Railroad run four times daily. During the watering-season, carriages run regularly four times a day between Great Barrington and the springs.

As embodying the best information at command in reference to this spring, I insert the following extract from a letter from Dr. C. T. Collins to Dr. Valentine Mott, for which I am indebted to Dr. John Bell's recent volume on the "Mineral and Thermal Springs of the United States and Canada:"—

"I must not close this letter without mentioning a very valuable mineral spring, situated among the mountains, a short distance from this village, and which has for many years past had a high *local* reputation for the cure of scrofula and eruptive diseases of the skin.

"The people in this part of the country consider it a specific for the cure of all that class of eruptive diseases which are popularly called by the vague and indefinite term of *salt-rheum*.

"During the past year, by way of experiment, I have placed several obstinate cases of eczema, ecthyma, acne, porrigo, etc. under the exclusive treatment of this water, and the results have been very satisfactory. Indeed, I may say that, in some cases, its effect was most extraordinary. So pleased was I with the use of this mineral water that I sent a jug of it to New York City, and had it analyzed by Professor Dorémus and Dr. Blake, the former assistant of Professor Silliman. It was found to contain soda, chlorine, carbonic acid, and a trace of alumina. Yet there is but little taste in it other than that of pure water. When bathed in, it imparts to the skin the most delightful softness of any that I have ever used, causing even a rough skin to feel smooth."

Arrangements exist here for the comfortable use of warm, cold, and shower baths.

CHAPTER XLI.

SPRINGS OF NEW JERSEY AND MAINE — SCHOOLEY'S MOUNTAIN.

The principal watering-place in New Jersey is *Schooley's Mountain Spring*, situated in Morris County, nineteen miles northwest from Morristown, and fifty from the city of New York. The water of this spring finds its exit from the earth near the summit of Schooley's Mountain, whence it is conveyed some distance down the mountain to a platform for the use of visitors, as a beverage and a bath. The quantity flowing from the spring is uniformly about thirty gallons in an hour. Its temperature is 50° Fahr. Its taste is strongly chalybeate, and it deposits oxide of iron readily upon substances with which it comes in contact. Its source is in the neighborhood of beds of iron ore, some of which, on both sides of the mountain, are worked advantageously in furnaces.

The waters of this spring have been known to possess valuable medicinal properties for more than three-quarters of a century, and for this reason, as well as on account of the salubrious atmosphere and its picturesque and romantic scenery, Schooley's Mountain has long been celebrated as one of the most desirable summer resorts for health and pleasure.

According to a chemical examination of the water by Dr. Nevin, its chief ingredients are "muriate and sulphate of lime and carbonated oxide of iron."

Dr. Bell remarks that "as a pure carbonated chalybeate, the water of Schooley's Mountain Spring is well adapted to a variety of maladies marked chiefly by anæmia, debility, and mucous discharges in which there is no inflammation of an organ present. Its tendency to induce constipation must be watched, and this effect arrested by the use of mild aperients."

Visitors to the springs from New York will go to Morristown by railroad and thence by stage, or to the White House by railroad and thence by stage. The springs are reached from Philadelphia by way of New Brunswick, and thence by stage, six miles, to Bound Brook, on the New Jersey Central Railroad. By this route they reach the White House, and thence, by stage, the springs.

SPRINGS OF MAINE.

Dr. C. P. Jackson, in a report upon the Geology of Maine, gives some account of two mineral springs in this State, the Saline Spring of Lubec, and Dexter's Chalybeate Spring.

The SALINE LUBEC SPRING rises near the junction of the blue limestone and red sandstone rocks, on the banks of a small stream near the head of Lubec Bay. He represents the water as clear and colorless, with a specific gravity of 1.025. The solid residuum of an Imperial gallon, perfectly dry, was 322.5 grains; 100 grains of this dry salt gave, by analysis, in one pint of water, the following results:—

	Grains.	Grains.
Chloride of sodium	64.0	199.000
Sulphate of lime	3.6	11.210
Chloride of magnesium	20.2	62.840
Sulphate of soda	9.0	27.985
Carbonate of iron	0.8	2.490
Carbonate of lime	2.0	6.250
Chloride of calcium	a trace.	12.720 loss.
Carbonic acid gas.		
	99.6	322.500
	.4 loss.	
	100.0	

DEXTER CHALYBEATE SPRING is located on the eastern branch of a stream known as Sebasticook. It deposits copiously "an ochreous yellow oxide of iron." Dr. Jackson considers this water a valuable tonic in various disorders of the digestive functions.

CHAPTER XLII.

MINERAL AND THERMAL WATERS BETWEEN THE MISSISSIPPI AND THE PACIFIC OCEAN.

In California—Oregon—Kansas—New Mexico—Wyoming—Utah, etc.

I DEPART from my general plan of treating only such springs as are improved for public use, to notice, in a brief way, the principal thermal and mineral fountains that have been discovered in the vast regions extending from the western borders of Iowa, Missouri, and Arkansas to the Pacific Ocean.

In the States of California, Oregon, Nevada, and Kansas, as well as in the Territories of Idaho, New Mexico, Wyoming, Colorado, Utah, etc., mineral and thermal waters are found in large abundance, of very positive quality, and of high temperature.

In NORTH OR UPPER CALIFORNIA, west of the Cascade Range, and at the foot of *Shasta Peak*, springs are found *hot enough, as travelers tell us, to boil eggs*. The region around is volcanic, and the bare summit of the Peak, rising to a height of from 12,000 to 14,000 feet, is regarded as an extinct volcano.

A few miles distant from the spring just mentioned is an *acidulo-chalybeate* fountain, and so sparkling, pungent, and effervescent is it that the trappers call it *Soda Water*.

Dr. Le Conte describes a number of *volcanic springs* in the Desert of Colorado, in Southern California, some of which are said to resemble the mud volcanoes of

Taman, in the Crimea, and others the eruptive springs or geysers in Iceland. They are in the neighborhood, and but six or eight miles distant from a range of volcanic hills from 800 to 1000 feet high. These springs consist of "numerous circular lakes, containing boiling mud, and exhaling a naphtha-like odor. Many of them are incrusted with inspissated mud, forming cones three to four feet high, from the apex of which proceed mingled vapors of water, sal-ammoniac, and sulphur. Four of them eject steam and clear saline water, with great violence, resembling in appearance the jet from the pipe of a high-pressure engine." These springs are in a muddy plain, bordering on a saline lake.

A hot sulphur spring, of the temperature of 137° Fahr., exists near Warner's Rancheria, about ninety miles from the Colorado, in South California.

IDAHO furnishes numerous mineral and thermal springs of very decided character.

The *Beer Springs*, described by Colonel Fremont, are about 135 miles, in a direct line, from the South Pass, through the Wind River Mountains, which separate the waters that flow into the Atlantic from those that find their way into the Pacific.

The *Beer or Soda Springs* are *carbonated waters*. They are described by Colonel Fremont as existing in great abundance in an amphitheatre of mineral waters, which is inclosed by the mountains that sweep around the circular bend of Bear River at its most northern point in the Territory of Idaho.

In the immediate neighborhood of the Beer or Soda Springs Colonel Fremont discovered a very remarkable fountain, which throws up its waters in the form of a *jet d'eau* to a variable height of about three feet. The flow of the water is accompanied by a "subterranean noise, which, together with the motion of the water, makes very much the impression of a steamboat in motion," and hence it was named the *Steamboat Spring*. This is a carburetted water of the temperature of 87° Fahr. "Within, perhaps, two yards of the *jet d'eau* is a small hole of about an inch in diameter, through which, at regular intervals, escapes a blast of hot air, with a light wreath of smoke, accompanied by a regular noise."

Hot Springs.—About two hundred and thirty miles northwest from Fort Hall are found hot springs of the temperature of 164° Fahr.

OREGON has numerous thermal springs, of which we mention the following:—

Malheur River Springs.—At the distance of one hundred and twenty miles in a northwestern direction from the Hot Springs of Idaho, mentioned above, are the Malheur Hot Springs. They are in latitude 44° 17' N., and longitude 117° W. Their temperature is 193° Fahr. Elevation above the sea, 1880 feet.

Hot and Warm Springs of Falls River.—These springs are on both sides of Falls River, in latitude 44° 40' N., 121° 5' W. longitude. They are about two hundred miles west from the Malheur River Springs.

The most noted springs of COLORADO are *the Carburetted or Boiling Springs of Pike's Peak.* On the Southern route from Independence, in Missouri, to Oregon and California, the traveler passes the now famous Pike's Peak, at the foot of which, and ten miles from Puebla, are found the Boiling Springs. Their elevation is 6350 feet above the ocean; their latitude 38° 42' north.

Colonel Fremont describes these springs as numerous, and some of them as unique and very beautiful. He says, "I came suddenly upon a large, smooth rock, about twenty yards in diameter, where the water from several springs was bubbling and boiling up in the midst of a white incrustation with which it had covered a portion of the rock." In describing one of this group, he says, "In the upper part of the rock, which had apparently been formed by deposition, was a beautiful white basin, overhung by currant-bushes, in which the cold, clear water bubbled up, in constant motion by the escaping gas, and overflowing the rock, which it had almost entirely covered with a smooth crust of glistening white."

These waters belong to the *acidulous class*, and are highly carburetted. They are said much to resemble the waters of the famous Seltzer Springs in the duchy of Nassau. Their temperature is variable, ranging, under different circumstances of the atmosphere, from 54° to 69° Fahrenheit.

NEW MEXICO has numerous mineral and thermal springs, some of which are sulphurous, but they have not been described with sufficient accuracy to make us acquainted either with their peculiar characteristics or their precise localities.

There are several springs in WYOMING that have attracted the attention of scientific travelers. Both Colonel Fremont and Captain Stansbury, in their respective narratives, notice the

FORT LARAMIE SPRING.—This fountain, thermal in its character, is ten miles from Fort Laramie, between the North Fork of the Platte and the Laramie Rivers, in latitude 42° 15′ N., and longitude 104° 47′ W. It is in the southeastern portion of the Territory, 625 miles from St. Joseph's, in Missouri. Its temperature is 74° Fahr., about the same as the Sweet Springs in Virginia.

In the western part of Wyoming, and in the midst of the *Salt Plains*, in the valley of the Sweet Water River, are found what are known as the *Ponds of Saleratus*. The chief of these ponds appeared to Captain Stansbury "as if frozen over, and covered with a light coating of driven snow. It was found to be a slight depression, about 400 yards long by 150 in width, covered with an effervescence of carbonate of soda, left by the evaporation of the water which had held it in solution." This substance is quite abundant, and emigrants use it in their culinary operations in preference to the saleratus of the shops.

Hot Springs of Pyramid Lake, Nevada.—The Pyramid Lake, embosomed in the Sierra Nevada Mountains, with its singular pyramidal mount, rising from its transparent waters to the height of about 600 feet, and walled in by almost perpendicular precipices, in some places nearly 3000 feet high, is a remarkable formation, and is said to have nothing to resemble it in any other portion of the world. Its boiling springs have attracted the attention of the scientific. Colonel Fremont describes them in about 39° N. latitude, and 117° 30' W. longitude, as boiling up with much noise. He states that the largest basin is several hundred feet in circumference, and has a circular space at one end of 15 feet in diameter, entirely filled with boiling water, whose temperature near the edge is from 206° to 208° Fahr. Its depth, near the centre, is more than 16 feet. The water is impregnated with common salt, but not so much so as to render it unfit for general cooking, and a mixture of snow makes it pleasant to drink.

The late Captain Gunnison, speaking of these springs, says, "At the base of the hills, around the lake, issue numerous *warm springs*, that collect in pools and smaller lakes, inviting aquatic fowl, during the winter, to resort to their agreeable temperature, and where insect larvæ furnish food at all times, and the soil is so heated that snow cannot lie in the vicinity. In some places springs of different temperature are in close proximity; some so hot that the hand cannot be thrust in them without pain."

Utah Territory, more than any other portion of North America, abounds in thermal waters, many of which are sulphurous and saline, and of very high temperature.

CITY WARM SULPHUR SPRINGS issue from a mountain on the immediate confines of Salt Lake City, and its waters are conveyed by pipes into bathing-houses, within the city, for the use of the inhabitants. The water is sulphurous, and yields, upon analysis, the carbonates of lime and magnesia, with small portions of the chlorides of calcium and sodium, together with sulphate of soda.

Three miles distant, and rising from the side of the mountain just mentioned, another spring flows out with great boldness. The temperature of its water is 128° Fahr. The specific gravity of this water is very slightly greater than that of distilled water. It contains chloride of sodium and traces of chlorides of calcium and magnesium, sulphate and carbonate of lime and silica.

Between Salt Lake City and the Great Salt Lake there are numerous *warm fountains*, which, Captain Gunnison informs us, deposit gypsum and other sulphates. They constitute delightful bathing, but are said to destroy the fertility of the soil to which their waters are applied.

Colonel Fremont thus describes a group of *hot springs* situated thirty-four miles north of Salt Lake City:—"In about seven miles from Clear Creek, the trail brought us to a place at the foot of the mountain, where there issued, with considerable force, *ten or twelve* hot springs, highly impregnated with salt. In one of them the thermometer stood at 136°, and in another at 132° Fahr., and the water, which spread in pools over the low grounds, was colored red." His analysis of this red earthy matter showed it to be highly impregnated with iron, and to contain the carbonates of magnesia and lime, with sulphate of lime, chloride of sodium, with silica and alumina.

Near *Bear River* is a depression, in which issue three fountains between the strata, within the space of thirty feet, of which one is *hot sulphur*, the next *tepid and salt*, and the other cool, delicious drinking-water. The three currents unite, and flow off through the plain, forming the beginning of a large and bold river.

Water of the Great Salt Lake.—Dr. Gale, of Washington City, has examined the water of this wonderful saline reservoir. He describes it as perfectly clear, with a specific gravity of 1.170; common water being 1.000. One hundred parts evaporated to dryness gave 22.422 of solid contents, consisting of chloride of sodium 20.196, sulphate of soda 1.834, chloride of magnesium 0.252, with a trace of chloride of calcium. Dr. G. regards this water as the purest and most concentrated brine in the world. The strongest salines of the Syracuse wells in New York contain but 17.35 per cent of the chloride of sodium.

Various *salt* and *sulphur springs* arise from the mountains and plains near the Great Salt Lake, and flow into it.

Thermal Saline Springs.—Captain Stansbury, in his narrative, informs us of the *Warm Saline*, whose temperature is 74° Fahr., that breaks out from the mountain at the northern end of the lake, and of the *Warm Springs* in the same locality, whose temperature is 84° Fahr.

We are told that the whole western shore of Salt Lake, bounded by an immense plain of soft mud, is traversed by numerous rills of sulphurous and salt water, that mostly sink into the earth, or are evaporated before they reach the lake.

Thermal Saline Springs of Spring Valley.—In this valley, lying on the western side of the mountain that extends in a southerly direction from the south end of Salt Lake, thermal saline springs are so numerous as to give the name to their location. Their temperature is generally about 74° Fahr.

TABLE EXHIBITING THE THERMALIZATION OF THE VARIOUS WARM AND HOT SPRINGS OF THE UNITED STATES AND ITS TERRITORIES.

I have thought that it would be interesting to my readers to have a condensed view of the various *thermal* springs of the United States and its Territories.

Virginia is rich in thermal waters, and up to the time of the discovery of the numerous hot springs of New Mexico, was regarded as possessing more of this class of waters than any other portion of the continent.

I shall first notice the thermal waters of Virginia and West Virginia, and shall regard all the springs as belonging to that class whose waters are distinctly above the mean temperature of the immediate country in which they arise. In this class I include the Greenbrier White Sulphur, although not generally regarded as a thermal spring; but the fact that it is full ten degrees above the mean temperature of the atmosphere and the media through which it flows, as well as of the neighboring fountains, properly gives to it that character.

	Fahrenheit.
White Sulphur, West Virginia	62°
Holston Springs, Scott County, Virginia	68°
Bath, Berkeley County, West Virginia	73°
Sweet Springs, Monroe County, West Virginia	73 to 74°
Red Sweet, Alleghany County, Virginia	75 to 79°
Healing Spring, Bath County, Virginia	85°
Warm Springs, Bath County, Virginia	98°
Hot Springs, Bath County, Virginia	98 to 106°
Perry County, Pennsylvania	72°
Lebanon, New York	73°
Merriwether County, Georgia	95°
Buncombe County, North Carolina	94 to 104°
Warm Springs, French Broad, Tennessee	95°
Florida Sulphur Springs	70°
Washita, Arkansas	140 to 156°
Spring near Fort Laramie, Wyoming	74°
Hot Sulphur Springs of California	137°
Hot Springs at Shasta Peak, California	.....
Great Salt Lake City Warm Springs	.....
Great Salt Lake Hot Springs, Utah	123°

	Fahrenheit.
Great Salt Lake Hot Chalybeate, thirty miles from Great Salt Lake..	132 to 136°
Great Salt Lake Thermal Saline............................	74 to 84°
Great Salt Lake Spring Valley Saline.....................	70 to 74°
Bear River Warm and Hot Springs, seventy-four miles northwest from Salt Lake City............................	134°
Lake Utah Warm Springs.......................................	.....
Hot Springs, Idaho...	164°
Malheur River Hot Springs, Oregon.......................	193°
Hot and Warm Springs, Falls River, Oregon..........	89 to 134°
Hot Springs, Pyramid Lake, Nevada*.....................	206 to 208°

* Mineral and Thermal Springs of the United States, by Bell.

CHAPTER XLIII.

MINERAL SPRINGS OF CANADA.

THE CALEDONIA SPRINGS.—These springs are situated about forty miles from Montreal, and a few miles south of the Ottawa River. They are a place of considerable resort during the summer season. There are four springs in this group deserving of notice. They are known as the *Gas*, the *Saline*, the *Sulphur*, and the *Intermitting Spring*.

The first three issue through a pliocene clay, within a few rods of each other. They are all more or less alkaline in character, the *Sulphur* the most so. The intermitting spring is two miles distant from the others, abounds in earthy chlorides, and emits carburetted hydrogen gas largely at irregular intervals.

1. THE GAS SPRING.—The temperature of this spring was found to be $44.4°$ when the thermometer stood in the air at $61.7°$. It discharges about four gallons of water per minute, and evolves a gas, ascertained to be carburetted hydrogen, at the rate of 300 cubic inches a minute. Its specific gravity is 1006.2; its taste pleasantly saline, without bitterness; its saline ingredients in 1000 parts, 7.7775. Carbonic acid in 100 cubic inches, 17.5.

2. SALINE SPRING.—This spring is not very dissimilar from the one just named, but, notwithstanding, from the name it bears, is somewhat *less* saline. Its temperature and specific gravity are essentially the

same. Occasionally it emits a stray bubble of carburetted hydrogen, but the amount of that gas evolved is very small. It is somewhat more strongly alkaline than the Gas Spring. This spring yields 10 gallons per minute, and to every 1000 parts of its water gives 7.347 parts of solid matter. Its free carbonic acid is 14.7 cubic inches in 100 cubic inches of water.

3. SULPHUR SPRING.—The water of this spring is slightly sulphurous in taste and odor. Solid matter in 1000 parts, 4.9506. It is somewhat more alkaline than the other springs of the group, contains silica in a relatively large proportion, and exhibits traces of iodine and iron.

4. INTERMITTING SPRING.—The temperature of this spring was 50° when the atmosphere around was 61°. Solid matter in 1000 parts of its waters, 14.639 parts. Chemical examination detects the existence of bromine, chlorine, and iodine in the water, with sodium, potassium, magnesium, and calcium. A large portion of the two latter exist in the form of chlorides. Traces of alumina and iron are also found.

TUSCARORA ACID SPRING.—This spring is located in Tuscarora Township, 21 miles north of Port Dover. Its waters abound in free sulphuric acid, in the proportion of 4 parts in 1000, and, also, with the sulphate of the alkalies, magnesia, lime, alumina, and iron in small quantities. It emits occasional bubbles of carburetted hydrogen, and its waters are acid and styptic to the taste, and decidedly sulphurous, while the odor of sulphuretted hydrogen is manifest for some distance around the spring.

CHARLOTTESVILLE SULPHUR SPRING.—This spring is in the neighborhood of Port Dover, on Lake Erie. Its waters are sparkling and limpid, their odor strongly sulphurous. The taste of the water is pungent, with a slight impression of sweetness, leaving a sense of warmth in the mouth. Chemical examinations show the presence of chlorides and sulphates in the water; the bases are ascertained to be soda, potash, magnesia, and lime, with traces of iron and alumina. It abounds very strongly in sulphuretted hydrogen, containing 26.8 cubic inches to the gallon. Its solid matter is 2.49446 parts to 1000.

MINERAL ARTESIAN WELLS *at St. Catharine's, Ontario.*—The analysis of this water, as reported in a printed circular, is very extraordinary. If the published statement of its analysis, by Dr. Chilton, be correct, and the water sent to him for examination was the natural water of St. Catharine's, the quantities in which its ingredients are held in solution, when we consider their peculiar character, are unexampled in the history of mineral fountains.

Dr. John Bell,[*] with amiable manifestations of incredulity, remarks, "Assuming the printed statements of the results of an analysis, by Dr. James R. Chilton, to be correct, the saline ingredients of this water are in a singularly large proportion, and this, too, of certain salts which are far from being common, still less abundant, in mineral springs. A pint of the water is represented to hold in solution 5064.15 grains of saline substances, which are equal to nearly five-sevenths of the watery menstruum in which they are dissolved. In other words, 16 ounces of the water hold in solution rather more than 10½ ounces of saline matter. They are in the following proportions in one pint of water; its specific gravity at 60° Fahr. being 1.0347:—

Chloride of calcium	2950.40
Chloride of magnesium	1289.76
Chloride of sodium	781.36
Protochloride of iron	13.76
Sulphate of lime	16.32

---

[*] Mineral and Thermal Waters of the United States and Canada.

Carbonates of lime and magnesia	2.08
Bromide of magnesium	a trace.
Iodide of magnesium	a trace.
Silica and alumina	.47
Grains	5064.15

"According to this analysis, the proportion of chloride of calcium (muriate of lime) in the water is a little more even than that which is found in the solution of this salt directed by the Pharmacopœia of the United States, viz., one part of the chloride in two and a half parts of the solution." On reading a little further, after the table of constituents of this water, we come to a "Card to the Public," in which we learn that the product of the artesian well is subjected to a certain process of depuration and evaporation, and that "that part which is composed of common salt first settles and is removed ; the remainder is dipped into vats until the earthy matter subsides, and then bottled off without any drug or admixture whatever being added thereto." Dr. Bell adds, "One thing seems to be certain, that the water bottled and sent away is a water prepared from that of St. Catharine's well, but not the water the direct flow from the vein or veins 'opened by boring.'" He further adds, in proof of the wonderful differences in the strength of the saline impregnations of different specimens of this water, that Mr. J. E. Young, an intelligent chemist, examined a specimen of this water left at the shop of Professor Procter, of the Philadelphia College of Pharmacy, with the assurance that it was from St. Catharine's well, in its original state, with the following results:—"Specific gravity, 1.390; saline contents in one ounce, 164 grains, and in one pint, 2624 grains. This last, large as is the proportion, is only a little more than one-half of the quantity of the salts contained in a pint of the water sent to Dr. Chilton for analysis."

VARENNES SPRINGS.—These springs are on the St.

Lawrence, seventeen miles below Montreal. Many years ago they were largely resorted to, but less so of late years, though probably from no want of merit in the waters.

There are two springs here, called the *Gas* and the *Saline Spring*. Both springs contain iodide, chloride, and bromide of sodium, with carbonates of soda, strontia, baryta, lime, magnesia, and iron. The temperature of the water is 45° to 47° Fahr.

St. Leon Spring is a *saline chalybeate*, similar in its general character to the springs of Varennes, but containing more iron. It emits large quantities of carburetted hydrogen gas.

The Plantagenet Spring derives its name from the township in which it is situated. It is near the river Ottawa. It resembles in the general character of its waters the St. Leon Spring.

Caxton Spring.—This spring is found in Caxton Township, on the river Yarnachiche. It resembles very much the St. Leon and Plantagenet Springs in the character of its waters, and, like the St. Leon, evolves large quantities of carburetted hydrogen.

INDEX.

	PAGE
Acid Springs, New York	249
Adams County Springs, Ohio	189
Adirondack Springs, New York	254
Administration—Remarks, etc.	33
Albany Artesian Wells, New York	236
Alburgh Springs, Vermont	270
Alleghany Springs, Virginia	165
Allison's Springs, Tennessee	196
Alterative Effects of Mineral Waters	29–83
Alum Springs, Rogersville, Tennessee	199
Analysis White Sulphur Water	67
Ancient Use of Mineral Waters	21
Avon Springs, New York	240
Bailey's Springs, Alabama	209
Ballston Springs, New York	227
Bath Alum Springs, Virginia	147
Bath Springs, Pennsylvania	265
Bedford Springs, Tennessee	255
Beersheba Springs, Tennessee	196
Berkeley Springs, Virginia	159
Berkshire Springs, Massachusetts	272
Best Time for Visiting Springs	57
Bethesda Springs, Wisconsin	194
Bladen Springs, Alabama	208
Blue Lick Springs, Kentucky	186
Blue Ridge Springs, Virginia	164
Buffalo Springs, Virginia	178
Byron Acid Springs, New York	253
Caledonia Springs, Pennsylvania	266
California, Springs of	277
Canada, Springs of	286–290
Capon Springs, Virginia	161
Carlisle Spring, Pennsylvania	264
Catoosa Springs, Georgia	207
Catskill Spring, New York	248
Chalybeate Spring at White Sulphur	108

INDEX.

	PAGE
Chalybeate Spring near Pittsburg	263
Changing from Spring to Spring	42
Chappaqua Spring, New York	247
Charleston Artesian Well	205
Chick's Spring, South Carolina	205
Chittenong Springs, New York	245
Clifton Springs, New York	245
Cold Sulphur Spring, Virginia	150
Columbian Spring, Saratoga	225
Congress Spring, New York	222
Cooper's Well, Mississippi	210
Corner's Black and White Sulphur, Virginia	163
Diet and Exercise at Springs	45
Directions for Use of Saratoga Waters, and Diseases for which used	232–236
Directions for Use of White Sulphur	81
Diseases for which White Sulphur should not be used	107
Diseases treated by White Sulphur	91–107
Doubling Gap Spring, Pennsylvania	264
Dress at Mineral Springs	44
Empire Spring, Saratoga	225
Ephrata Springs, Pennsylvania	266
Errors and Abuses in the Use of Mineral Waters	37, 230–232
Estill Springs, Kentucky	187
Experience the only Sure Guide	23
Fauquier White Sulphur, Virginia	178
Fayette Springs, Pennsylvania	265
Flat Rock, Saratoga	225
Florida, Springs of	218
Frankfort Springs, Pennsylvania	262
French Lick Springs, Indiana	190
Gettysburg Springs, Pennsylvania	258
Geyser, or Spouting, Saratoga	226
Glenn's Springs, South Carolina	204
Gordon's Springs, Georgia	207
Grayson White Sulphur, Virginia	174
Halleck's Spring, New York	236
Hamilton Spring, Saratoga	225
Harrodsburg Springs, Kentucky	183
Harrowgate Springs, New York	247
Healing Springs, Virginia	137
High Rock Spring, Saratoga	223
Holston Springs, Virginia	175
Hot Springs, Arkansas	214
Hot Springs, Bath County, Virginia	128
Huguenot Springs, Virginia	180
Iodine or Walton Spring, Saratoga	225
Iodine Springs, Georgia	206
Johnson's or Hollins's Institute, Virginia	164
Jones's White Sulphur, North Carolina	202

INDEX.

	PAGE
Jordon Rockbridge Alum, Virginia	146
Jordon's White Sulphur, Virginia	157
Kittrell's Springs, North Carolina	203
Lebanon Springs, New York	253
Lee's Springs, Tennessee	198
Length of Time to use Mineral Waters	31
Liability to Mistakes as to Sulphur Waters	40
Madison Springs, Georgia	206
Maine, Springs of	275
Massanetta Springs, Virginia	156
Medical Advice deemed essential in Europe, etc	37, 38, 230
Medicinal Efficacy of Mineral Waters	26
Medicines with Mineral Waters	50
Messina Springs, New York	246
Mineral Waters not a Catholicon	26–37
Missisquoi Springs, Vermont	269
Modus Operandi of Mineral Waters	28
Montgomery White Sulphur, Virginia	170
Montvale Springs, Tennessee	197
Newberry Springs, New York	147
Newburg Springs, Vermont	269
New London Alum Springs, Virginia	181
New River White Sulphur, Virginia	114
New York Springs	219–254
Ocean Springs, Mississippi	212
Ohio White Sulphur, Ohio	188
Olympian Springs, Kentucky	185
Oregon, Springs of	278
Pavilion Spring, Saratoga	223
Periods for the Use of Mineral Waters	47
Periods of the Year for Visiting Springs	57
Perry County Spring, Pennsylvania	264
Preparations for Use of White Sulphur, etc	84
Prescribing Mineral Waters	50–55, 229
Pulaski Alum Springs, Virginia	174
Pulse, Effects of White Sulphur, etc	86
Putnam Spring, Saratoga	223
Rawley Springs, Virginia	154
Red Sulphur Springs, Virginia	111
Reed's Springs, New York	236
Resemblance of some Mineral Waters to Mercury in their Effects	35
Richfield Springs, New York	244
Roanoke Red Sulphur, Virginia	164
Robertson's Springs, Tennessee	196
Rochester Springs, Kentucky	184
Rochester Springs, New York	248
Rockbridge Alum Springs, Virginia	141
Rockbridge Baths, Virginia	150
Routes to West Virginia and Virginia Springs	61
Saline and Gaseous Efficacy of White Sulphur Waters	71

INDEX.

	PAGE
Salivation from Sulphur Waters	36
Salt Sulphur Springs, Virginia	109
Saratoga Alum, Saratoga	225
Saratoga Waters, how to be used, etc.	234
Schooley's Mountain Springs, New Jersey	274
Sharon Springs, New York	274
Shocco Springs, North Carolina	202
Silk important as a Dress	45
Springs in New Mexico, etc.	277–283
St. Louis Magnetic Springs, Michigan	192
Stribling's Springs, Virginia	151
Sweet Chalybeate, or Red Sweet, Virginia	121
Sweet Springs, Virginia	115
Synopsis of Important Facts in the Use of White Sulphur Water	87
Tallahatta Springs, Alabama	209
Tate's Springs, Tennessee	198
Thermalization of Mineral Waters	131
Thermalization Table of Mineral Waters	284
Union Springs, Saratoga	223
Variety Springs, Virginia	151
Vermont Springs, Vermont	269
Verona Springs, New York	248
Virginia and West Virginia Springs	59
Warm and Hot Bathing, Cautions, etc.	201
Warm and Hot Springs, North Carolina	200
Warm Springs, Bath County, Virginia	134
Warm Springs, French Broad, Tennessee	199
Warm Springs, Georgia	206
Westport Springs, Ohio	190
West's Spring, South Carolina	204
White Creek Springs, Tennessee	196
White Sulphur Springs, North Carolina	203
White Sulphur Springs, West Virginia	62
Williamstown Spring, South Carolina	205
Winchester Springs, Tennessee	196
Yellow Springs, Ohio	189
Yellow Springs, Pennsylvania	266
Yellow Sulphur Springs, Virginia	171
York Springs, Pennsylvania	263

INDEX TO CARDS.

Coleman & Rogers' Pharmacy and Mineral Water Depot,
Baltimore 3
Gettysburg Springs, Pennsylvania 4–6
Hollins Institute, Virginia 7
Jordon Rockbridge Alum Springs, Virginia . . . 8
Hot Springs of Arkansas 9
Levy Brothers, Merchants, Richmond . . . 11
Massanetta Springs, Virginia 12
Montvale Springs, Tennessee 13
Piedmont & Arlington Life Insurance Co., Richmond . 14
Rawley Springs, Virginia 15
Roanoke College, Virginia 16
Stieff's Pianos 17
Sweet Chalybeate Springs, Virginia 18
Wade & Boykin, Druggists, etc., Baltimore . . 19
Warm Springs, Virginia 20
White Sulphur Springs, West Virginia . . . 21
Yellow Sulphur Springs, Virginia 22

COLEMAN & ROGERS'
Pharmacy and Mineral Water
DEPOT,
178 W. BALTIMORE ST.,
BALTIMORE, MD.

OUR STOCK EMBRACES A FULL LINE OF

PURE DRUGS

AND

Rare PHARMACEUTICAL PREPARATIONS,

CHEMICALS, Etc.

ALSO,

NATURAL MINERAL WATERS

FROM THE

MOST POPULAR MEDICINAL SPRINGS

IN

France, Germany, and the United States.

Cash Orders solicited, and satisfaction guaranteed.

THE
Gettysburg Katalysine Water.

Dr. John Bell, author of a standard medical work on Mineral Springs, says of it:—

"The Gettysburg Water has produced signally curative and restorative effects in different forms of Dyspepsia, Sickness of the Stomach, Heart-burn, Water-brash, Acute Neuralgic Pains, Loss of Appetite, Chronic Diarrhœa, Torpid Liver, Gout, Chronic Rheumatism, Nodosities of the Joints, Approaching and Actual Paralysis, Diabetes, Kidney Disease, Gravel," etc.

Dr. J. J. Moorman, resident Physician at the White Sulphur Springs, Professor of the Washington Medical University at Baltimore, and author of one of the best works on the use of Mineral Waters, writes:—

"That as a solvent of the uretic concretions in rheumatism and gout, it promises to take a high rank among the medicinal springs of Europe and America. This solvent power is not claimed, that I am aware of, in behalf of any other mineral water or medical agent."

The New York *Medical Record* editorially says:—

"We have also seen cases of albuminuria much relieved by it, as well as the irritable bladder of old age and calculous disorders of the lithic acid diathesis. From experiments made on our own person, as well as others, we can state that the Gettysburg Water is a regulator of all the secretions and excretions: under its influence the kidneys and liver, the glands of the intestinal canal and the skin, all perform their normal functions; the bowels, if constipated, become regular; the skin, if dry, becomes moist; the torpid liver is excited to healthy action, and the kidneys perform their functions with perfect regularity. There is a total absence of any disagreeable sensation whatever; the vis medicatrix seems roused to increased activity, and all morbid causes of bodily or even mental disorder seem rapidly to pass away. The result is—increased appetite and digestion, freer circulation, a stronger pulse, a calmer mind, a more tranquil sleep, a clearer complexion, and an increasing nervous and muscular power. . . . Where gouty or rheumatic persons are taking the Water, we find an extraordinary quantity of uric acid secreted or deposited from the urine; the sweat no longer contains this principle in excess, as it generally does in gouty subjects; and with proper attention to regimen and diet, the health rapidly improves, distorted limbs become straightened, and enlarged joints gradually reduced to their natural size."

For further reports from the medical profession, and of wonderful cures, send for pamphlets.

WHITNEY BROS., General Agents,
227 SOUTH FRONT ST., PHILADELPHIA.

THE GETTYSBURG KATALYSINE SPRING

is situated near the historic town of Gettysburg, Adams County, Pa. By an interesting coincidence it appears on the spot over which was fired the first gun in the great and decisive battle of our late War of Rebellion fought at this place. All around is historic ground. Though a local tradition ascribes to this Spring healing power, it was not generally considered medicinal until after the battle. The rumor that some of the wounded combatants had received benefit from the use of its waters, current at Gettysburg after the battle, induced resort to it by invalids, with results which can hardly find a parallel in the medical history of the world. The establishment of a great Spa had previously been the work of centuries; but the Gettysburg Katalysine Spring leaped, by a single bound, from obscurity to the foremost rank among modern medical sources. The thirteenth revised edition of the United States Dispensatory classes this American Spring of yesterday with the renowned Carbonate Spas of the Old World, the Vichy, and the Pyrmont, while nearly every newspaper and medical journal of America has chronicled some of its wonderful cures.

The enterprise, resulting in the erection of a *large hotel* at this Spring, was suggested by the published correspondence of Governors Curtin and Geary, and of General Meade, which commended it as eminently national and philanthropic. It was afterwards indorsed by the subjoined appeal of the members of the National Congress :—

"The undersigned, deeply impressed with the wonderful curative prodigy which appears on the battle-field of Gettysburg, and learning that it is the design of public-spirited citizens to utilize this great discovery in the cause of medical science, and in the interest of humanity, by erecting in the vicinity a hotel for the entertainment of the afflicted of our own and other countries who may seek here their lost health, and of the patriotic pilgrims to these holy grounds, deem it to be our duty to commend the proposed enterprise as eminently philanthropic and praiseworthy.

ULYSSES MERCUR,
W. P. FESSENDEN,
E. D. MORGAN,
ROSCOE CONKLING,
T. A. PLANTS,
E. R. ECKLEY,
JOHN TRIMBLE,
N. B. JUDD,
H. I. DAWES,
WM. MOORE,
H. W. CORBETT,
B. F. RICE,
J. A. GARFIELD,
JOHN A. LOGAN,
ORANGE FERRISS,
J. G. BLAINE,
B. F. HOPKINS,
C. D. HUBBARD,
W. B. WASHBURN,
B. F. WADE,
SCHUYLER COLFAX,
ORRIS S. FERRY,
HENRY WILSON,
L. M. MORRILL,
JOHN COVODE,
OAKES AMES,
GODLOVE S. ORTH,
JOHN COBURN,
JOHN TAFFE,
W. G. COFFIN,
R. R. BUTLER,
W. B. STOKES,
J. W. McCLURG,
CHAS. UPSON,
WM. H. KOONTZ,
GEO. LAURENCE,
F. C. BEAMAN,
J. F. BENJAMIN,
JOHN HILL,
H. D. WASHBURN,
H. L. CAKE,
ALEX. RAMSEY,
GEO. H. WILLIAMS,
MORTON C. HUNTER,
W. MUNGEN,
D. A. NUNN,
T. D. ELIOT,
DANL. J. MORRELL,
W. H. HOOPER,
AMASA COBB,
B. F. LOAN,

E. EGGLESTON,
C. A. NEWCOMB,
W. WILLIAMS,
H. B. ANTHONY,
J. B. HENDERSON,
J. M. HOWARD,
M. WELKER,
W. SPRAGUE,
A. H. LAFLIN,
H. VAN AERNAM,
GEO. P. VAN WYCK,
WM. HIGBY,
T. W. FERRY,
LUKE P. POLAND,
F. E. TROWBRIDGE,
WM. A. PILE,
GEO. W. ANDERSON,
WM. LOUGHRIDGE,
J. J. GRAVELY,
RUFUS MALLORY,
R. P. BUCKLAND,
JAS. S. MARVIN,
F. STONE,
RICHARD YATES,
HORACE MAYNARD,
A. H. BAILEY,
R. W. CLARKE,
GEO. W. JULIAN,
CHAS. O'NEILL,
BURT VAN HORN,
WM. M. STEWART,
GEO. M. ADAMS,
W. H. KELSEY,
B. M. BOYER,
CHAS. SITGREAVES,
JOHN BEATTY,
G. F. MILLER,
C. T. HULBURD,
WM. D. KELLEY,
J. T. WILSON,
J. K. MOOREHEAD,
GEO. A. HALSEY,
T. VAN HORN,
JAMES W. NYE,
F. T. FRELINGHUYSEN,
GEORGE VICKERS,
W. S. LINCOLN,
JAMES M. CAVANAUGH,
SIDNEY CLARKE,
JACOB BENTON.

HOLLINS INSTITUTE,

Botetourt Springs,
ROANOKE COUNTY, VIRGINIA.

COL. GEORGE P. TALOE,
President of Trustees.

CHARLES L. COCKE, A.M.,
General Superintendent.

BOARD OF INSTRUCTION AND GOVERNMENT.
SESSION 1872-73.

JOSEPH A. TURNER, M.A., Modern Languages, Ethics, and English.
CHARLES L. COCKE, A.M., Mathematics and Chapel Exercises.
MISS BETTIE D. FOWLKES, Painting, Drawing, and Mathematics.
MADAME A. BUTTEL, Colloquial French and German.
MISS JULIA PORCHER, Instrumental Music and Vocalization.
MISS SALLY BROWNE RYLAND, Preparatory School.
MRS. SUSANNA V. COCKE, Domestic Department.
WM. H. PLEASANT, Ancient Languages, History, and Science.
AUGUST BUTTEL, Director of Music Department and Piano.
MISS SALLY L. COCKE, Languages and English.
MISS ROSA P. COCKE, Languages and History.
MISS CYNTHIA McGAVOCK, Instrumental Music and Singing.
MRS. MARY E. SLOAN, Superintendent of Music-Rooms.
MRS. H. R. McVEIGH, Matron.
MRS. FANNY THOMAS, Matron.

In this Institute there are nine Departments of Instruction:—I. English Language and Literature. II. Ancient Languages and Literature. III. Modern Languages and Literature (French and German). IV. Mathematics. V. Natural Sciences. VI. Mental and Moral Science. VII. History. VIII. Music. IX. Drawing and Painting.

The Institute is well provided with Musical Instruments, including fifteen Pianos, Organ, etc., Chemical and Philosophical Apparatus, Minerals, Maps, etc. Sessions open about the 15th of September, and continue *nine* months. Pupils may come in at any season of the year, and remain throughout the period of their school-days, including vacations. *Parents of pupils are boarded during summer* at moderate cost.

This place, formerly known as "*Johnston's Springs,*" has not been kept as a public "*Watering-Place*" for thirty years, the premises having been wholly devoted to school purposes. It is, however, a delightful *Summer Residence*, enjoying the advantages of mineral waters, and a few orderly people are received as private boarders during summer.

☞ POST-OFFICE, Botetourt Springs, Va.
DEPOT, Salem, Va. & Tenn. R.R.

JORDON ROCKBRIDGE
ALUM SPRINGS,
Rockbridge County, Virginia.

These Springs are 8 miles from *Goshen Depot*, on the Chesapeake and Ohio Railroad, from which point *Coaches* run regularly over a good road to and from the Springs in connection with the Cars.

The improvements here are entirely new, and embrace the modern conveniences for comfortable accommodations.

The principal Hotel, in addition to *Parlors, Dining-Room, Ball-Room*, etc., contains upwards of one hundred *Chambers*, all newly furnished, for the accommodation of families or individuals. There are also *Cottage* accommodations outside the Hotel.

☞ In addition to the *Alum Waters*, whose medicinal waters are too well known to make it necessary to speak of them here, there is on the grounds one of the strongest and best *Chalybeate Springs* of the country, which, as a *direct* and *powerful tonic*, is well deserving the attention of Spring visitors. There is also within visiting distance from the Hotel another Spring, known as "*Iodine* and *Alum Water*," which possesses valuable medicinal powers, and some peculiar to itself, and which will be constantly kept fresh at the Hotel for the use of visitors. The waters of this Spring are not only adapted to the cure of the various diseases commonly cured by Alum Waters, but also, from its peculiar and highly *Alterative* composition, to be a reliable remedy in other cases wherein these waters are uncertain or inefficient.

☞ Facilities for *Recreation* and *Amusement* usually found at fashionable Watering-Places will be found here.

☞ *Post-Office*, known as "*Alum Springs*," is kept in the Hotel.

☞ During the season an *Office of the Western Union Telegraph Co.* will also be kept in the Hotel, communicating with all parts of the world.

LEVY BROTHERS.

THE LARGEST
DRY GOODS HOUSE
IN THE STATE,

Nos. 1017 and 1019 Main Street,
RICHMOND, VA.

TERMS CASH.
ONE PRICE AND NO DEVIATION.

Purchasing their goods direct from the importers, manufacturers, and at the auction trade sales, enable them to offer extra inducements to purchasers of dry goods.

Prompt attention given to orders.

For particulars read daily papers published in Richmond, Petersburg, and Lynchburg.

MASSANETTA SPRINGS,

ROCKINGHAM COUNTY, VA.

The Water from the *Ague* or *Taylor's* Spring, taken at the Spring, has had historic fame for very many years for especial and specific powers in all chronic diseases originating in malaria, such as *Ague* and *Fever*, *Enlargements* of the *Liver* and *Spleen*, Chronic Inflammations of the *Bronchia*, *Stomach*, *Kidneys*, *Bowels*, *Bladder*, etc. etc., originating in Ague and Fever, Yellow, Congestive, and Bilious Fevers.

The Water is soft, mucilaginous, and a more safe, pleasant, and sure remedy at the Spring than Quinine. *It bears shipment well.* We believe it to be antidotal to Miasmatic Poison; and, if drunk in any swamp or miasmatic locality, a preventive of malarial diseases.

For Dyspepsia, the Rachitic and Cachectic diseases of Children; in Scrofula, Diphtheria, Scurvy, and in Womb and Venereal diseases, the combined Ulcer and Ague Water is a charming remedy.

For cost, carriage, etc. of transported Waters, address

B. CHRISMAN, President,
COWAN'S STATION,
ROCKINGHAM COUNTY, VA.

MONTVALE SPRINGS,

Blount County, East Tennessee.

This favorite Summer Resort is 25 miles south of *Knoxville*, in a sequestered valley, almost encircled by lofty spurs of the "*Chilhowee*" Mountain, which here embosom a valley of surpassing loveliness, in which these Springs have their source. Their elevation is 1400 feet above the level of the sea.

The remarkable power of these Waters in the cure of functional derangements of the *Liver, Bowels, Kidneys, and Skin*, and indeed of *Chronic Diseases* generally, fully attests their high medicinal properties, and has long made them a place of large public resort.

☞ All the *accessories* for *Recreation* and *Amusement* usually found at fashionable Watering-Places will be found here.

☞ ROUTE.—Visitors to *Montvale* will necessarily pass over the *East Tennessee and Virginia* or the *Georgia Railroad*, making the city of *Knoxville* a point; thence by way of the Knoxville and Charleston Railroad to *Marysville*, 16 miles; from which place they are conveyed in Coaches, running in connection with the Trains, to the Springs, 9 miles distant.

☞ The Springs will be open for the reception of Visitors on the 15th of May, and kept in a style worthy of the patronage of a discriminating public.

☞ For *Pamphlets* containing *Analysis* and general description of the Waters, address

JOSEPH L. KING,
Montvale Springs, East Tennessee.

PIEDMONT AND ARLINGTON LIFE INSURANCE COMPANY,

HOME OFFICE,

RICHMOND, VA.

W. C. CARRINGTON, *President.*
J. E. EDWARDS, *Vice-President.*
Prof. E. B. SMITH, *Actuary.*

D. J. HARTSOOK, *Secretary.*
J. J. HOPKINS, *Assistant Secretary.*
B. C. HARTSOOK, *Cashier.*

Annual Income over One and a Quarter Million Dollars.

Policies Liberal and Non-Forfeitable. Losses below the Average of other like Companies.

All approved and thoroughly tested forms of Life and Endowment Policies issued.

This Company has conducted its business at a SMALLER RATIO OF EXPENSE TO INCOME than any other Company of same age in America.

Just and liberal dealings with all its policy-holders, promptness in paying claims, and the special advantages it presents to patrons, have secured to the Company unequaled success, and guarantee its continued prosperity.

Surplus divided annually among policy-holders. Retiring policy-holders dealt with liberally.

$100,000 deposited with Treasurer of Virginia, and in other States, for additional security of policy-holders.

The Company has complied with the requirements of the Insurance Departments of New York, Ohio, California, Kentucky, etc. No other Southern Life Company has established itself in New York.

NO COMPANY CAN OFFER SUPERIOR ADVANTAGES.

Over 18,700 Policies issued to March 1, 1873.

RAWLEY SPRINGS,

ROCKINGHAM COUNTY,

11 MILES FROM HARRISONBURG, VA.

We announce to the Spring-going public that these Springs, so long and favorably known for their efficacy in the treatment of a large circle of diseases, will be open for the Season of 1873 on

THE FIRST DAY OF JUNE.

These Waters have long been regarded as the *strongest* and *most fortunately compounded* Waters, that are *distinctly chalybeate* in character,—the union of other valuable medicinal ingredients with the iron making them not only *actually Tonic*, but also highly *Alterative* in their effects.

☞ The usual facilities for *Amusement* and *Recreation* found at fashionable Watering-Places generally, will be found here.

Every proper effort will be made to make our guests comfortable, and to insure the continuance of the large patronage the Springs have heretofore enjoyed.

CHARGES FOR THE SEASON.

Board per month $60.00
"   " week 15.00
"   " day 2.50
Children and Servants, *half price.*

☞ *Rawley* may be reached conveniently from the *North* and *East* by the Manassas Gap Railroad to *Harrisonburg;* and from the *South* and *West*, from *Staunton*, via *Harrisonburg.*

☞ *Omnibuses* will run from the Springs daily, in connection with the Railroad Cars.

A. B. IRICH,
President Board of Directors.

J. N. WOODWARD, Superintendent.

April, 1873.

ROANOKE COLLEGE,

SALEM, VA.

Founded 1853.

The Annual Sessions commence on the First Wednesday in September, and close the Third Wednesday in June.

COURSE OF STUDY.

The thorough and comprehensive curriculum, extending over a period of four years, embraces the following:—Classical, Oriental, and Modern Languages, English Language, Belles-Lettres, History, and Literature, Moral and Intellectual Philosophy, Mathematics, Natural Sciences, International Law, Political Economy, with Lectures on Physiology and Hygiene.

LOCATION.

In point of location Roanoke College challenges comparison with any other institution in America. The Roanoke Valley, in which it is situated, is unsurpassed for its fertility, beautiful mountain scenery, equable temperature, general healthfulness, and freedom from malarious diseases. Salem, the most thriving town in Southwest Virginia, is immediately on the Atlantic, Mississippi and Ohio Railroad, and at the junction of the Valley Railroad, now under contract.

EXPENSES.

☞ The Expenses for a Session of TEN MONTHS (including Tuition, Board, Fuel, Lights, Washing, etc.) are ABOUT $200. A slight advance on this estimate must be made for students boarding in private families. Full details given in the annual catalogue.

☞ The *low price of board* ($10 to $14 per month), consequent upon the abundance of the country, enables this Institution to educate young men on more reasonable terms than are offered by any other Institution of *high grade* in the South.

☞ The unsurpassed advantages of Roanoke College have gained for it a wide and rapidly increasing popularity, students being in attendance annually from *fourteen* to *eighteen* States and Territories.

Persons desiring fuller information are referred to Dr. J. J. Moorman, Physician to White Sulphur Springs, and Lecturer on Physiology and Hygiene in Roanoke College.

For Catalogues and further particulars, address

Rev. D. F. BITTLE, D.D., Pres't.

STIEFF'S PIANOS.

Upwards of fifty First Premiums, Gold and Silver Medals, were awarded to Charles M. Stieff for the best Piano, in competition with all the leading manufacturers in the country.

Office and Warerooms, No. 9 N. Liberty Street, Baltimore, Md.

The superiority of the Unrivaled Stieff Piano-Forte is conceded by all who have compared it with others. In their New Grand Square Scale, seven and one-third Octaves, the manufacturer has succeeded in making the most perfect Piano-Forte possible.

Prices will be found as reasonable as consistent with thorough workmanship.

A large assortment of second-hand Pianos always on hand, from $75 to $300.

We are agents for the celebrated Burdett Cabinet, Parlor, and Church Organs, all styles and prices, to suit every one; guaranteed to be fully equal to any made.

Send for illustrated catalogue containing the names of over 1500 Southerners, 500 of whom are Virginians, 200 North Carolinians, 150 East Tennesseeans, and others throughout the South, who have bought the Stieff Piano since the close of the war.

Sweet Chalybeate Springs,

Formerly known as the RED SWEET SPRINGS,

Alleghany County, Va.

These Springs, so long and favorably known for their valuable *tonic* and *alterative* powers, both as a BEVERAGE and BATH, have been newly and completely refitted, with convenient and comfortable accommodations for 400 persons.

Their situation is central in the *Great Spring Region*, being 16 miles south of the *White Sulphur*, and 9 miles from *Alleghany* Station, on the *Chesapeake and Ohio Railroad*.

They will be open for the Reception of Visitors, for the

Season of 1873, on the 1st day of June.

The Various Sources of Recreation and Amusement,

common to the best-conducted Watering-Places, will be kept up for the accommodation of Visitors. Parties suffering from *Dyspepsia, Neuralgia, Chronic Diarrhœa, Spermatorrhœa, Fluor Albus, Amenorrhœa, General Debility*, and especially *Incipient Consumption*, will find the most decided and beneficial results following the use of these Waters.

☞ Valuable as these Waters are admitted to be when used as a *Beverage*, the great charm of the place, pleasurably, hygienically, and medicinally, is found in the large *Inclosed Pools for Plunge Bathing*, and in the well-arranged SHOWER and TUB BATHS of any degree of *temperature* that may be desired.

Taking the *Bathing* facilities here, all and in all, they are believed to be equal, or superior, to any elsewhere to be found in the country.

☞ Arrangements have been made for the residence at the Springs, during the Season, of a highly-competent Physician.

JOHN KELLY, Proprietor.

WADE & BOYKIN,
No. 3 LIBERTY ST.,
BALTIMORE,

IMPORTERS AND WHOLESALE DEALERS IN

Drugs, Paints, Oils, Chemicals,
Etc. Etc.

In addition to calling the attention of *Dealers* and *Physicians* to our carefully selected stock in our current business, we desire to elicit the attention of the general public to Dr. Wade's

Liver Corrector and Dyspepsia Cure.

Dr. Wade, having used this medicine with great success in his private practice for many years, has been induced to allow it to be put up under his especial care for general use in the diseases for which he has successfully prescribed it.

We confidently recommend a trial of this remedy to those who are afflicted with diseases of the *Liver* or *Stomach*, or with *Constipation of the Bowels*, for the cure of which it is a prompt, safe, and reliable remedy.

It is *purely Vegetable* in composition, and free from all alcoholic admixture. It has been successfully employed by many of the leading citizens of this and other States.

In addition to the diseases above mentioned, this remedy has been very successfully used for the relief of *Sick Headache*, *Jaundice*, and in biliary conditions of the system generally.

For sale by Druggists generally.

WARM SPRINGS,
BATH COUNTY, VA.

This famous *Spa*, long distinguished for its *luxurious Bathing facilities*, and for the cure of many diseases properly treated by Warm Bathing, is open for the reception of Visitors.

Among the diseases for the cure of which these Waters have long been distinguished, we mention *Atonic Gout*, *Rheumatism*, *Lymphatic* enlargements, *Paralysis*, *Obstructions of the Liver and Spleen*, *Syphiloid* affections, *Cutaneous* diseases, *Nephritic* and *Calculous* disorders, and the various chronic forms of *Female Obstructions*.

The facilities here for efficient and pleasant POOL BATHING are not surpassed in America. The arrangement of *Pools* and *Dressing-Rooms exclusively for Ladies* commands universal approval.

☞ The facilities for *Amusements* usually found at fashionable Watering-Places.

☞ These Springs are distant from Millboro', on the Chesapeake and Ohio Railroad, 15 miles; from Covington, on the same road, 22 miles. From both places, pleasant and safe *Coaches* run over good roads in connection with the Railroad Cars.

☞ Travelers from the *North* should leave the Cars at *Millboro'*. Those from the *West*, at *Covington* or *Millboro'*.

☞ *Telegraphic* Office in Hotel.

☞ An experienced Physician resides at the place.

☞ PAMPHLETS forwarded, by mail, on application.

JOHN L. EUBANK,
Acting Partner and Superintendent.

WHITE SULPHUR SPRINGS,

Greenbrier County, W. Va.

The undersigned beg leave to announce that these Springs, so long celebrated for their valuable ALTERATIVE WATERS, their charming summer climate, and the large and fashionable crowds that annually resort to them, will be open for the Season of 1873 on the

15th DAY OF MAY.

Their capacity for accommodation is from 1500 to 2000 persons.

☞ Prof. Rosenberger's celebrated BAND will be in attendance to enliven the *Lawns* and *Ball-Room*.

☞ *Masquerades* and *Fancy Balls* as usual through the Season.

☞ An extensive LIVERY for the use of Visitors.

HOT AND WARM SULPHUR BATHS,

so efficacious in many cases, always at the command of the Visitor.

☞ The *Chesapeake and Ohio Railroad* is now in excellent running order to the Springs both from the *East* and *West*.

☞ A *Telegraph Line* is in operation to the Springs.

CHARGES FOR THE SEASON.

Board per day	$3.00
" " month of 30 days	80.00

Children and Colored Servants, half price.
White Servants, according to accommodations furnished.

☞ We have the pleasure to announce to those who design to visit the Springs, that Prof. J. J. MOORMAN, M.D., well known as the author of several valuable books on MINERAL WATERS, and for 35 *years* the PHYSICIAN TO THE WHITE SULPHUR, will be at the Springs this summer in that capacity.

GEO. L. PEYTON & CO.

WHITE SULPHUR SPRINGS, W. VA., *March*, 1873.

YELLOW SULPHUR SPRINGS,

Near Christiansburg,

Montgomery County, Va.

These Springs, so long distinguished for their active *Tonic* and valuable *Alterative powers*, will be opened for the Season of 1873 ON THE FIRST DAY OF JUNE.

☞ The facilities for *Amusement* and *Recreation* usually found at first-class Watering-Places will be found here.

☞ *Telegraphic* and *Express* lines are in operation to the Springs.

☞ *Hot and Warm Baths of the Mineral Water*, so essential to many invalids, at command of the Visitors.

Extensive additional improvements are now in progress, to be completed by the commencement of the season, among others, a large and commodious HOTEL *with all the modern improvements.*

The immense increase of visitation to this place within the last few years has made such extension of our improvements a necessity.

These Springs arise with great boldness near the summit of the Alleghany Mountain, more than 2000 feet above the level of the sea, the *most elevated and coolest summer resort in Virginia;* the climate being as salubrious, and the air as elastic and invigorating, as can well be imagined.

As an efficient *Tonic*, this water has maintained an unsurpassed reputation for seventy years. As an *Alterative* in many chronic affections, it has proved a blessing to thousands.

Owing to its fine tonic and alterative powers, its therapeutic applicabilities are extensive, but especially has it exhibited its curative powers in *Dyspepsia*, and chronic affections of the *Abdominal Cavity;* in *General Debility* and *Nervous Prostration.* In various chronic affections of the *Skin*, in *Kidney* disorders, and in *Chlorosis* and kindred *female affections*, it has had a very large success.

☞ *For Terms*, which will be moderate, see our *Pamphlet*, which will be sent on application.

☞ *Excursion Tickets* to the place can be obtained at all the principal Railroad Offices.

J. J. & J. WADE, Proprietors.

YELLOW SULPHUR SPRINGS, VA., *March*, 1873.

www.ingramcontent.com/pod-product-compliance
Lightning Source LLC
Chambersburg PA
CBHW030741230426
43667CB00007B/800